Scientists

and the Development
of Nuclear Weapons

THE CONTROL OF NATURE
Series Editors: Margaret C. Jacob and
Spencer R. Weart

PUBLISHED
SCIENTISTS AND THE DEVELOPMENT OF NUCLEAR WEAPONS
From Fission to the Limited Test Ban Treaty, 1939-1963
Lawrence Badash

EINSTEIN AND OUR WORLD
David Cassidy

NEWTON AND THE CULTURE OF NEWTONIANISM
Betty Jo Teeter Dobbs and Margaret C. Jacob

FORTHCOMING
CONTROLLING HUMAN HEREDITY
1865 to the Present
Diane B. Paul

CONTROL OF NATURE

Scientists
and the Development
of Nuclear Weapons

From Fission to the
Limited Test Ban Treaty
1939-1963
Lawrence Badash

HB

Humanity Books

An Imprint of Prometheus Books
59 John Glenn Drive
Amherst, New York 14228-2119

First published in 1995 by Humanities Press International, Inc.
165 First Avenue, Atlantic Highlands, New Jersey 07716

Inquiries should be addressed to
Prometheus Books
59 John Glenn Drive
Amherst, New York 14228–2119
VOICE: 716–691–0133, ext. 210
FAX: 716–691–0137
WWW.PROMETHEUSBOOKS.COM
12 11 10   8 7 6

Library of Congress Cataloging-in-Information Data
Badash, Lawrence.
     Scientists and the development of nuclear weapons : from fission
to the Limited Test Ban Treaty, 1939–1963 / Lawrence Badash.
          p.   cm. — (The Control of Nature)
     Includes bibliographical references and index.
     ISBN 13: 978-1-57392-538-9
     1. Nuclear weapons—History.   2. Arms race—History.   I. Title.
II. Series.
     U264.B34   1995
     355.02′17′09—dc20   94–17471
                                          CIP

A catalog record for this book is available from the British Library.

Printed in the United States of America on acid-free paper

Contents

Series Editors' Preface

THIS NEW SERIES of historical studies aims to enrich the understanding of the role that science and technology have played in the history of Western civilization and culture, and through that in the emerging modern world civilization. Each author has written with students and general readers, not specialists, in mind, and the volumes have been written by scholars distinguished in their particular fields. The author of this book, Lawrence Badash, is known for his work and teaching on subjects ranging from the early history of radioactivity to the Cold War.

The aim of this book on the advent of the "atomic age" is not to just lay out some basic historical information, which could only be a sample of the many complex developments that scholars are currently exploring. Still more this volume intends to show the chief questions and debates that engage current historical scholarship.

The current debates as presented here emphasize the "Control of Nature." While not excluding a discussion of how knowledge itself develops, how it is constructed through the interplay of research into nature with the values and beliefs of the researcher, this volume—like all the others in the series—looks primarily at how science and technology interact with economic, social, linguistic, and intellectual life, in ways that transform the relationship between human beings and nature. In every volume we are asking the student to think about how the modern world came to be invented, a world where the call for progress and the need to respect humanity and nature produce a tension, on the one hand liberating, on the other threatening to overwhelm human resources and ingenuity. The scientists whom you will meet here could not in every case have foreseen the kind of power that modern science and technology now offer. But they were also dreamers and doers—as well as shrewd promoters—who changed forever the way people view the natural world.

Finally, a word about citation style: whole books and essays are cited within brackets, e.g., as [1], referring to items in each bibliography. When a page or volume and page are cited, the notation includes a colon, as [1:32] or [1:vol. 2, 13]; when many items in the bibliography are cited, each is separated by a semicolon, as [1; 2; 3].

MARGARET C. JACOB
SPENCER WEART

Acknowledgements

Sincere thanks are due to Bruce Badash, Leonard Marsak, and Spencer Weart for their help and advice. Thanks are due to the fine production editors at Humanities Press, Stacey Anolick and Kathy Delfosse.

Introduction

T HE COLD WAR may be over, but its legacy is not. There are tens of thousands of nuclear weapons in the world, national economies are drained of treasure, international relations among many countries reflect the period's tensions, and the psychological stress upon generations of people around the globe, who anticipated nuclear annihilation at one time or another, has taken its toll. In short, though the period may have come to an end, it was a time of major historical significance, still worthy of our attention.

The bombing of Hiroshima and Nagasaki with newly invented nuclear weapons in August 1945 did not merely herald the imminent end of World War II. It opened a Pandora's box of other issues, problems, and opportunities. Was this a "moral" weapon, appropriate for a principled nation to use against cities and their civilian inhabitants? Was war finally too destructive to fight? (This question had been asked upon the introduction of previous "ultimate" weapons, such as gunpowder, the rifle, ironclad ships, the machine gun, the tank, aircraft, and poison gas, only to have them included in the arsenals of the day—and used.) Could one defend against the awesome power of these new weapons? Would potentially hostile nations be able to acquire them? What would be the role of science in the nation's future, and how should this goose that laid the golden egg be fed? Can any technical problem be solved, if enough energy and resources are provided? With American leadership, could a new, peaceful world order be established, holding the atomic club as a threat to recalcitrant states? Would large standing armies, and their large budgets, go the way of the horse cavalry, leading to a leaner, cheaper military? The questions were endless, and few are resolved even today.

Victory in 1945, unlike the conclusion of most other wars, did not bring expected peace and stability. Indeed, the United States and its allies soon were engaged in a Cold War with the Soviet Union and its partners. A permanent war economy in the United States kept a new entity, the military-industrial complex, well nourished. And amid the weapon systems galore, pride of place was held by the warheads and

1

delivery systems termed nuclear. The consequences are debatable. Vast amounts of treasure were spent by many nations until the Cold War ended in the late 1980s. (Military budgets may since have fallen somewhat, but conventional weapons are not inexpensive.) And, despite numerous calls for their employment, nuclear weapons were not used in any conflict after World War II. But was nuclear war avoided because of our arsenal, or despite it? Would verifiable arms control have been a safer and far less costly approach to international relations? Was the level of Cold War tension and hostility exacerbated by the nuclear bombs and warheads targeted on the enemy and on us?

This short book will attempt to present such questions in the context of the period; answers, however, will largely be left to the readers, for these are so often matters of opinion. This is not a conventional history of the nuclear arms race, for the role of science and the contributions of scientists—both their discoveries and political activities—are stressed. At the start of our story, scientists had limited interaction with, and virtually no support from, most governments. By its end, scientists were America's fair-haired boys (and girls) who made miracles happen— except when they were feared as security risks. In between, the nature of the relationship between science and government changed: politicians found that science was too important to be left to the scientists, and scientists learned that he who pays the piper calls the tune. Rather than begin our story in 1939, with the discovery of nuclear fission, a better perspective will be gained if we turn first to an earlier period, when several trends were set in motion.

Ruminations About Science

Power over Nature

WITH CONTEMPORARY AIR conditioning and heating, massive dams to control water supplies, use of pesticides, medical life-support systems, smog from automobile exhausts, and destruction of rain forests, among many other things, we take for granted humanity's ability to alter if not control aspects of its environment. Such power did not always exist; indeed, the concept of mastery over nature seems not to have been given much thought in antiquity, and only in the Middle Ages did it slowly mature. Finally, by the time of the Scientific Revolution of the seventeenth century—the age of Copernicus, Kepler, Galileo, Descartes, Leibnitz, Newton, and others—the view that the goal of scientific investigation was to oversee nature could be clearly expressed by the famous philosopher and England's lord chancellor, Francis Bacon.

In his *Novum Organum* (1620), he asserted that "the true and lawful goal of the sciences is none other than this: that human life be endowed with new discoveries and powers." [8:56] Even though the biblical story of creation has Adam naming the animals, thereby assuming something of a ruling position, Bacon was careful to explain that humans' new-found authority through scientific investigation meant not subjugation of nature but accountable dominion over it: "nor can nature be commanded except by being obeyed." [6:22–23] This implied responsibilities, and sometimes, he wrote, it was proper to keep discoveries secret from the state. [7:190] Bacon's influence upon the Age of Reason was profound. Not only did people for the first time recognize that they could take charge of their own destinies, but the political philosopher John Locke expanded the concept to include political destiny: people could aspire to govern themselves.

Over the next centuries, with the Industrial Revolution and the growth of industrial societies, humans acquired increased power but largely ignored their responsibilities. As historian Lynn White phrased it, the vision changed from humans being one with nature to humans as exploiters of nature; indeed people came to believe "nature has no reason for existence save to serve man." [115:1207] This attitude, combined with the ferociousness of modern technology toward nature, spelled disaster. In the United States in the 1960s, motivated by such works as Rachel Carson's *Silent Spring*, a poetic reproach to our use of the pesticide DDT, the Environmental Protection Agency was established. The EPA symbolically marked public recognition of an environmental crisis and a nascent political will to deal with society's problems. It was a time, too, of a counterculture that rejected what it saw as the excesses and evils of industrialization and engaged in a search for alternative paths. [90] Western civilization, it was argued, need not follow only the aggressively distorted side of Bacon's philosophy; White pointed out that our traditions embrace also St. Francis of Assisi's call for humans to live in harmony with all other creatures. [115]

That knowledge is power is seen nowhere more clearly than in the development of nuclear weapons. That those who produce the knowledge have responsibilities (including Bacon's admonition to withhold it from those who would misuse it) manifested itself in the concept of science and social responsibility, and in the efforts of many scientists to further arms control negotiations. [26; 32] But while nuclear weapons burst suddenly upon the public's consciousness in 1945, the problems they raised were not entirely novel.

Big Science

The effort in the United States during World War II to make nuclear weapons, popularly called the Manhattan Project, was notable not least for the enormous size of everything but the atom's nucleus itself. The scale of the production plants, the amount of materials, the number of workers, the cost—all were described in superlatives. While other wartime projects also consumed comparable funding and personnel (e.g., penicillin production), the Manhattan Project easily was the model cited when one spoke of the new age of "Big Science." Before the war, most scientific research was conducted by individuals, using relatively small apparatus made in their laboratories, spending what little money could be scrounged from the few sources available or found in their own pockets. A valid picture comes from Cambridge University's Cavendish Laboratory, in the 1930s the world's premier physics institution: when a research student asked the shop steward for a piece of one-inch diameter steel pipe, he was given a hacksaw and directed to an abandoned bicycle in the courtyard. This was still the age of impecunious "little science," in which most experiments were conducted with as much cheap "sealing wax and string" as possible.

Big Science means research teams (comprised sometimes of dozens of people), with members contributing their specialized expertise; few scientists could master the range of skills required in some fields. The apparatus was necessarily larger than before and often built (at least in part) by engineering companies, the most striking examples being the particle accelerators of nuclear physics, such as the two-mile-long Stanford Linear Accelerator Center (SLAC, built in the 1960s) and the $8.2 billion Superconducting Super Collider (SSC), with a circumference of more than 50 miles, begun in Waxahachie, Texas, but killed in 1993 by a cost-cutting Congress. The source of funds? The government, of course, for only Washington has the resources for such endeavors (even if it sometimes changes its mind).

But Big Science did not burst full grown upon the scene in 1945. Like most activities, it was a process, not an event. We may, in fact, look back to the early nineteenth century, when the Herschels in England and then Lord Rosse in Ireland built larger and larger reflecting telescopes. Here, the instruments themselves were big, even if staffs and outside funding were not. This trend culminated in true Big Science in the 200-inch reflector atop Mount Palomar in California, whose mirror was completed after World War II. The pursuit of still greater diameters has now resumed on the summit of Hawaii's Mauna Kea. Wealthy foundations support such construction, while the government often underwrites the research. The observatories maintain permanent staffs to operate the facilities efficiently, and viewing time (a coveted commodity) is awarded by special committees to an international array of visitors.

Not only size but team research and government funding, too, may be traced to the nineteenth century, if not earlier. State and federal coast and geological surveys were a start of this tradition, while treks and voyages of exploration that included scientists also became more and more common. Notable among these ventures were the occasions when botanist Joseph Banks accompanied Royal Navy Captain James Cook on his first voyage to the South Seas (1769–71); a youthful Charles Darwin circumnavigated the globe as naturalist aboard the *Beagle* (1831–36); the Lewis and Clark expedition ventured across North America to the Pacific (1804–06), taking no scientists along but collecting much zoological, botanical, and ethnographic information nonetheless; the expedition of Captain Charles Wilkes, USN, to Alaska, Antarctica, and many Pacific islands (1838–42) corrected the previous omission of expertise by making scientists the mainstay of its personnel; and the famous British ship *Challenger* sailed on its oceanographic expedition

(1872–76), oriented even more toward science and less toward exploration.

The size of the undertaking and number of personnel of expeditions could be matched in some other fields of science, especially those that demanded cooperation. Thus, by the end of the nineteenth century, meteorology, epidemiology, seismology, weights and measures, and other subjects that could advance only by amassing regional or global data, or by international agreement on standards, were guided by international organizations.

The size of hardware grew too. Engineered apparatus—too big or complex to be constructed in the laboratory's workshop—increasingly made its appearance. Liquid air machines, so popular at the turn of the century for dramatic public demonstrations, such as the shattering of a cooled bouquet of flowers, allowed a great increase in low-temperature investigations. There were unique pieces also, constructed for particular studies. Peter Kapitza, working in Ernest Rutherford's Cavendish Laboratory in the late 1920s, had an industrial-size electric generator built for his experiments with very high magnetic fields. [14; 12] A few years later, also in the Cavendish, John Cockcroft and E. T. S. Walton built a machine to accelerate protons, the first successful "atom smasher," of which Ernest Lawrence's cyclotron in Berkeley, California, soon became the most popular design. [73]

While not every attribute of Big Science need be present in each case, the yardsticks of apparatus size, team research, cost, and government sponsorship were so common after 1945 that this way of doing science was firmly planted. But we must recognize that the roots had been growing for some time, and the construction of nuclear weapons, radar, the proximity fuse (which did not have to strike its target but detonated when close to it, thereby increasing the "kill radius"), and other wartime projects were evolutions of science's characteristics, not newly minted traits. [85]

Science and Government

Nor was government utilization of scientists entirely novel. Most scientists, to be sure, even those employed in state-supported universities, had virtually nothing to do with the corridors of power. But over the previous centuries, government had experienced needs or seized opportunities that required scientific talent. While under enlightened presidents, such as Thomas Jefferson and John Quincy Adams, technical projects might be proposed for their intellectual value, their justification to Congress inevitably rested upon more prosaic economic, health, or political arguments. Thus were established in the nineteenth century the Coast Survey to aid navigation, the Patent Office, numerous land surveys, the Department of Agriculture (one of whose functions was to help farmers improve the quantity and quality of their yields), the Geological Survey, the Weather Bureau, and other agencies that employed technically trained people.

The tradition accelerated in the twentieth century with the creation of the National Bureau of Standards, the Public Health Service, the Bureau of the Census, the Bureau of Mines, the National Institutes of Health, and still more vehicles to accomplish the government's bidding in specialized fields. And all this prior to World War II. [31] This background discussion points heavily toward the United States, this being the nation where the Manhattan Project occurred, but all nations, and industrialized nations in particular, have placed scientific talent on the payroll for a long, long time.

Government use of such expertise may be most obvious to the general public in wartime. The development and application of poison gases during World War I is a good example. While most people, including scientists, recoiled in horror at weapons that employed poisons, others argued that incapacitating agents were more humane than shells that blew humans to pieces. The Germans began the large-scale deployment of gas, but by the time of the Armistice British and American chemical plants were vastly outproducing them. Morality is a weak counter to patriotism, nationalism, the desire to "save our boys," and the argument of fewer deaths in the long run if a loathsome weapon is used in the short term. As with gas, the ethics of nuclear weapons would be debated periodically over the course of half a century.

Internationalism in Science

The customs of science changed as a result of World War II but, as we have seen, more in degree than in kind. Change occurred in yet another scientific fashion, that of the role of nationalism and internationalism in science, and it too has a history. For centuries, if not millennia, the language of scholarship in Western Christendom was Latin. With ease in communication, national borders were irrelevant in the spread of scientific ideas. By the seventeenth century, however, along with the rise of nation-states, use of the vernacular became increasingly common. Scientists now wrote in their national languages, but they continued to behave as if the study of nature were an international calling, and they seem to have convinced the rest of society that this was the proper viewpoint.

We find, for example, that in the mid-eighteenth century Harvard professor John Winthrop and his telescope were given safe passage across the battle lines of the French and Indian War. His respected goal was to observe the transit of Venus across the face of the sun from Newfoundland. [120] The message, of course, was that men of science were above the fray. But another message, not articulated only because its opposite was inconceivable at the time, was that science was irrelevant to national destinies.

About two decades later, in 1779, Benjamin Franklin, then an esteemed diplomat from the young American nation at the court of France, requested American naval vessels not to interfere with Captain Cook's voyage homeward from his exploration of the Pacific. Franklin's success in politics was based on his fame as one of the greatest scientists of his day, and, more than most others, he had experienced the value of the internationality of science first hand. The American Congress, less familiar with such civilized behavior, issued a countermanding order to intercept Captain Cook (who nevertheless made it home to England safely). [68:398–401] In another instance, in 1813 famed British chemist Humphrey Davy raised no eyebrows when he was allowed to travel to enemy France during the Napoleonic Wars to receive a medal for his electrical investigations. [69:603]

As the nineteenth century progressed, such personal experiences became more institutionalized. The number of scientific societies in many nations grew rapidly, and it became customary to salute distinguished foreigners with honorary academy membership and even medals. In certain fields international societies were created; their international congresses inevitably followed. Undergraduates often went abroad for their degrees, and postgraduates (called graduate students today) for research experience. And established scientists visited the laboratories of foreign colleagues, published frequently in foreign journals, corresponded with their peers irrespective of nationalism, and in general behaved as if the world of science were without political boundaries.

All this makes it sound as if scientists lived in a pleasant but unreal world, isolated in their ivy-covered towers from political and social currents of the day. For in actuality relations between nations were not always harmonious, and society in both North America and Europe underwent significant changes in the late nineteenth century. Indeed, and this is a point to remember, scientists may have been citizens of the world professionally, but they were personally citizens of various countries. The absent-minded-professor image of academics is a cartoon of the truth, for educators are far more likely to have multiple interests than people in most other occupations. The scholarly Adlai Stevenson, who twice ran for U.S. president against war hero Dwight Eisenhower, was ridiculed for being an "egghead," to which charge he replied: "With what other part of the anatomy would you prefer that one think?" Scientists were active participants in their communities and, like other citizens, embraced a range of likes, dislikes, hopes, and fears. In their laboratories they might be dispassionate and critical observers of the facts, but in the larger society, despite some claims to the contrary, they could be rabid nationalists. Scientists, after all, are human beings.

It will come as no surprise, then, to learn that priority disputes sometimes served as the rope in nationalistic tugs-of-war. The British and the Germans, for example, each claimed that a particular radiation law was discovered by citizens of their own lands. National pride embraced notable scientific contributions, much as nations today tally their Olympic medals or Nobel Prizes to certify their stature.

Real warfare inevitably was the occasion for stronger displays of chauvinism. Louis Pasteur protested the bombardment of Paris during the Franco-Prussian War by angrily returning to the University of Bonn the honorary degree it had awarded him. World War I provided the stage for still greater violations of the creed of internationalism, as scientists turned from bystanders at the hostilities into actors in them.

In an effort to show the purity of their patriotism, members of major British scientific societies, such as the Royal Institution, other Chemical Society, and the Royal Society, disparaged the loyalty of the members born in Germany or with German-sounding names or German accents. Despite longtime residence in Britain (in one case more than 50 years), service to these societies as officers, distinguished careers, and no evidence whatever of sympathy for Germany's goals, these men were for the most part hounded into relinquishing their positions and absenting themselves from the meetings.

Greater attention, of course, was directed toward enemy scientists and the quality of German science. Germany had long been considered a powerhouse in science, by many the preeminent nation in science; now some second thoughts were expressed. It was said that Germans exploited the discoveries of others; they were not themselves very innovative. (Americans echoed this sentiment toward Japan in the 1980s.) Germans were methodical and industrious but not original. Upon more serious consideration, however, the mastery of the German chemical industry was recognized, and the Allies worked hard to build up their own chemical facilities. Indeed, World War I came to be called a chemist's war.

The sentiments above, perhaps natural in the throes of wartime, might have remained generalized, and long-standing friendship toward individuals in the enemy camp might have persisted, but for a blunder. The folly was a manifesto, "To the Civilized World," which 93 German intellectuals (including 22 scientists and physicians) signed in October 1914. The document denied responsibility for the war, for invading neutral Belgium, and for harm to Belgian citizens and cities. In England, France, and America there was first amazement that intelligent people could so delude themselves about the facts, followed by anger at this support of German militarism. Although a number of the manifesto signers later said they had not seen the exact wording, or that they regarded it merely as patriotic support for their troops, their Entente colleagues took it to be a repudiation of the universality of science, for the signers were experienced in investigation and analysis of data. More than anything else, this manifesto was a touchstone for bitterness.

With surprisingly little curiosity about the specific contributions to warfare that German chemists were making (quite a different attitude would prevail in World War II), scientists in the Entente nations debated the wisdom of expelling their distinguished German colleagues from their societies. Should enemy aliens be allowed to continue as members? Did their scientific fame bring more distinction than condemnation? In the end, a variety of actions was taken. In retrospect, it seems that scientists spent as much energy during World War I discussing what they felt to be fitting and proper behavior as they did improving munitions, submarine detection, and poison gas. [10]

By the time of World War II, such moral distractions for scientists were minimal; they saw themselves engaged in a scientific race, and no one in the Allied camp could allow Hitler to acquire nuclear weapons first. Nor were there heated discussions about expelling Axis scientists from Allied societies; in fact, there was virtually no personal vituperation. Indeed, many refugee scientists, with foreign names and accents, became leading lights of the project to construct nuclear weapons in the United States. But in both wars, the revered belief in the internationalism of science suffered. Danish physicist Niels Bohr could note that "scientists have long considered themselves a brotherhood working in the service of common human ideas." [75:ix] Scientists like to view this as the normal condition. Twentieth-century warfare, however, brings different emotions as well as the needs of secrecy, and the contact, communication, and cooperation that are the hallmarks of typical internationalism in science become dormant for the duration.

Background to the Bomb

Refugees from Nazism

HITLER'S RISE TO power in 1933 was not the reason for anti-Semitism in Germany, but the laws his government passed legalized discriminatory acts against Jews. An exodus began that has been compared in importance to the flight of Christians from Constantinople, when the Turks seized that city in 1453 and the Byzantine Empire collapsed. For these earlier refugees carried with them manuscripts containing vital texts of Greek antiquity that were unknown to the Latin West or existed there only in translation from Arabic. This classical learning, including much science, helped to ignite the intellectual activity of the Renaissance. In like manner, the emigrés of the 1930s, particularly those learned in nuclear physics, helped to shift the "center of gravity" of world science from western Europe to the United States and played an invaluable role in the development of nuclear weapons.

In Germany, Jews were only 1 percent of the population. Yet in a university system that had but one professor in a department (other staff has lesser titles), Jews held more than 12 percent of the chairs. Additionally, Jews won one-quarter of the Nobel Prizes that had gone to Germany. Clearly, their achievements were out of proportion to their numbers, though their visibility made them better targets than they were already. Not long after the Nazi party took the reins of government, it issued a "Cleansing of the Civil Service" decree, aimed at removing "non-Aryans" from office, followed by other restrictive legislation. Since non-Aryans were defined as those having two or more Jewish grandparents, and since the universities were state institutions, these anti-Semitic acts forced many scholars from their positions. [22; 23; 24; 30]

By the start of World War II, almost 40 percent of university professors (numbering a few thousand) were dismissed, perhaps a third of them scientists. The list read like a "who's who" of learning. Albert Einstein,

11

easily the most famous scientist in the world, was the most prominent, but the roster included 16 others who already were, or who would become, Nobel laureates. To be sure, many distinguished scientists (presumably Aryans, and by no means all supporters of Hitler) remained in Germany, yet never in history has a country tried so hard to export its brains. Many of the refugees went to Great Britain; some to Scandinavia, Turkey, India, and other countries that welcomed them. Most came to the United States, where, despite the economic depression and a degree of domestic anti-Semitism, there were more colleges and universities relative to the size of the population, and hence more job opportunities. [24; 60; 112]

Scientists still in Germany faced a government whose ideology extended to science: aside from Nazism's perversion of biology to invent a "master race," those in power looked upon relativity and quantum mechanics as "Jewish physics," and thus not valid. Scientists abroad who hosted or otherwise helped the refugees, and the refugees themselves, were appalled by this political intrusion into science. All learned that science could no longer remain aloof from politics. A process of education began. Increasingly, scientists would fight to keep others from determining what they could study—for example, the response to Lysenko's attack on genetics in the Soviet Union. They would strive to protect the jobs of those fired for political reasons—for example, those who refused to sign the loyalty oaths imposed during the postwar McCarthy period in the United States. And they would articulate their views on the uses and misuses of science—for example, by suggesting means of verifying arms control treaties, and by criticizing the seemingly endless development of weapons for the arms race. Scientists had entered the political arena.

Radioactivity and Nuclear Physics

Nuclear weapons are often called atomic bombs, but the name is inaccurate. The energy that makes these explosives so awesome comes from reactions that take place not in the whole atom but in its tiny heart, the nucleus. A few words about the development of this science will help to place the story of the weapons in perspective.

The phenomenon of radioactivity was discovered in Paris in 1896 by the physicist Henri Becquerel. He performed experiments similar to those used to explore the recently discovered X-rays and detected a somewhat different type of radiation emitted from minerals that contained uranium. Soon he traced the activity to elemental uranium itself. It was a new process of nature, yet it could not compete for the attention of scientists with X-rays and several other kinds of radiation then being studied. In particular, fascinating pictures of the bones in a hand could be made more quickly and with greater sharpness by using X-rays. [16; 43]

Radioactivity gained the spotlight, however, two years later, after Marie Curie, a Polish student working in Paris, detected similar radiation from the element thorium. She was puzzled that her uranium ore was more active than its uranium content should allow. Tracing these "impurities" soon led to the discoveries of polonium and radium, both radioactive. Given the interesting properties of these substances, and the circumstance that new elements are not discovered every day, the lure of radioactivity was assured. [87]

One of those attracted to the new science was a young New Zealander doing postgraduate research in Cambridge University's Cavendish Laboratory. This was Ernest Rutherford, who became the central figure in radioactivity and nuclear physics once he turned his attention to the field. Rutherford studied radiation that ionized, that is, created charged particles as it went through a gas. He named the radiation that ionized strongly but was stopped by a sheet of paper "alpha," and the weaker ionizer but more penetrating radiation "beta." Another still more piercing radiation was found in 1900 by Paul Villard and named "gamma." [16]

There are numerous types of radiation. Some are composed of tiny particles, as are Rutherford's alpha and beta. Others are part of the electromagnetic spectrum, which embraces shortwave radio, VHF, UHF, infrared light, all colors of visible light, ultraviolet light, X-rays, and gamma rays. The turn of the century was a time of intense interest in many of these radiations. [15]

In 1898, Rutherford was given a physics professorship at McGill University in Montreal, where he built up a thriving research unit that focused mostly on radioactivity. Within a few years, during which he learned that thorium emits a gaseous radioactive product (which led to the recognition of series of decaying radioelements) and that the alpha ray is a positively charged particle of atomic dimensions (the beta had earlier been shown to be a negative electron), Rutherford, with chemist Frederick Soddy, explained what occurred in the phenomenon of radioactivity. It was, they said, modern alchemy; contradicting a few centuries of chemical belief, they argued that the atom was not stable. One element transmuted, or changed spontaneously into another, and that into yet another, in a series, until a stable product was reached. In each disintegration one or more types of radiation was emitted: an alpha or a beta, and often a gamma as well. Each active product had its characteristic half-life, the time for half its quantity to transform. Every radioelement (radioactive element) could thus be identified by its unique combination of half-life and radiation. This was a vital insight because the quantities of a number of the products in the different decay series were far too small for normal "wet" analytical chemistry identifications. [118]

In Paris at the turn of the century, Marie Curie and her physicist husband, Pierre, joined forces to investigate radioactivity. They recognized that the rays emitted from these bodies, especially from the powerful radium, carried much energy. These rays could turn glassware a light shade of purple, cause some crystals to glow, and inflict a burn upon skin. In an effort to quantify the energy, Pierre Curie and his student Albert Laborde found that a radium solution maintained a temperature above that of its surroundings and that the energy outflow was remarkably large. This was particularly interesting, for the nineteenth-century law of conservation of energy said that energy might be changed from one form to another (for example, heat to motion) but that "you can't get something for nothing." Yet radium was emitting prodigious quantities of energy without any apparent diminution (its half-life of 1,600 years made it impossible to notice the slow change in the number of its atoms). While the Curies speculated about some unknown etherial radiation that supplied the energy to radioactive bodies, other physicists argued that the energy was intrinsically atomic, buried there since the atom's creation at the beginning of the universe. The Rutherford-Soddy transmutation law gave the mechanism of its release: unstable atoms ejected alphas, betas, and sometimes gammas in an attempt to attain a stable condition. [16; 87]

After Rutherford moved to Manchester University in 1907, he proved that the alpha particle was a charged helium atom. Then, in 1911, based upon experiments in which alphas were deflected from their paths by their interactions with target atoms (like charges repelling each other), he proposed that the atom is mostly empty space, with almost all its mass concentrated in a tiny nucleus. It had been conjectured for some years that the atom had a structure, inasmuch as it could emit particles such as the alpha and beta. But now Rutherford brought experimental proof to bear and calculated that the nucleus was only about one ten-thousandth the size of the atom (10^{-12} and 10^{-8} centimeters, respectively). Soon the picture of negative electrons orbiting around the positive nucleus was accepted, and one may say that the subject of radioactivity evolved into nuclear physics. The nucleus, it became apparent, obeyed some laws of physics that were not familiar in everyday, human-size phenomena. [50; 118]

With most of his staff and students gone during World War I, and himself involved in war-related work, Rutherford had little time for his basic research. Nonetheless, he carried on a modest program that bore fruit just about the time the war ended and he accepted the position as director of the Cavendish Laboratory. When a small particle traveling very fast collides with something about the same size it can cause the "target" to move, as occurs with billiard balls. When Rutherford filled a glass tube with different types of gases and allowed alpha particles to strike the contents, he saw tiny flashes of light on a detector that was placed beyond the normal range of the alphas. This was logical, since the alphas pushed the gas particles some distance. But when the tube was filled with nitrogen, Rutherford observed far too many flashes, at unusually great ranges, and these flashes looked very much like those from hydrogen gas, with which he had become familiar in recent experiments.

What occurred, he reasoned, was a rupture of the nitrogen nucleus, with a hydrogen nucleus being one of the new particles formed. This nucleus (pictured then as a ball) was thought to be the smallest particle of matter having a single positive electrical charge, just as the electron was the smallest particle with one negative charge. (Together, of course, they form the electrically neutral hydrogen atom.) Soon called a proton, the hydrogen nucleus was regarded as a basic building block of all matter. As recognized several years before, radioactivity was the natural process of nuclear disintegration. Rutherford now showed that it was possible to produce disintegration artificially, using natural means: the alpha projectiles from normally decaying radioelements. [118]

In the early 1920s, Rutherford and his associates followed up on this discovery and planned to test more elements for their susceptibility to disintegration. They got no further than the lightest elements, however, for the electrical charges on the heavier nuclei repelled the alpha particle projectiles. It became clear that more energetic "bullets," and a more copious supply of them, would be necessary to attempt disintegration of the elements heavier than potassium. Rutherford began to talk of ten-million-volt apparatus that would impart such energy to projectiles, but the electrical industry was hard pressed to overcome the inadequate insulators of the day. Research students in the Cavendish, however, began to design and build devices to accelerate electrons and protons. Finally, in 1932, John Cockcroft and E. T. S. Walton succeeded in splitting lithium into two alphas when they fired energetic protons at it.

Repeating a point made above, radioactivity is the natural disintegration of nuclei, while Rutherford's work described in 1919 was artificial disintegration by natural means. Building upon this concept, the Cockcroft-Walton experiment was artificial disintegration by artificial means. Using well-insulated equipment to store large charges, Cockcroft and Walton accelerated protons across the voltage gap, giving them great velocity before they hit a lithium target. With a cloud chamber (a device for making the tracks of particles visible) attached to their accelerator, Cockcroft and Walton quickly were able to give the first experimental confirmation of the famous $E = mc^2$ relationship (energy equals mass times the square of the velocity of light), which Albert Einstein had proposed about a quarter of a century before. Their triumph invigorated several other accelerator projects, including Ernest Lawrence's earlier successful cyclotron, which was based on a novel way of speeding particles in a spiral path. [13]

In a well-known speech before the Royal Society in 1920, Rutherford predicted the existence of a particle with about the same small mass as a proton and no electrical charge—what would soon be named the "neutron." Although the experimental results upon which this forecast was based were incorrect, Rutherford's intuition that the neutron was a reality was valid. Through the early 1920s, he and his chief assistant, James Chadwick, searched unsuccessfully for this particle. It would be an ideal projectile: since it had no charge, it would not be repelled from a nucleus but could slip right in. In the early 1930s, however, news came first from two scientists in Germany, and then from others in France, that described an unusually penetrating "gamma ray." Chadwick disagreed with this interpretation, and in experiments in 1932 quickly showed that the "ray" was his long-sought neutron. [13]

The neutron was soon recognized as a constituent of the nucleus. Protons gave the nucleus its charge, which determined the chemical nature of the atom and so its place in the periodic table of elements. Neutrons provided the additional mass found in any given element.

Neutrons also figured in the obviously timely experimental investigation of encounters with the heavier elements. In this research, Enrico Fermi and his group in Rome led the way. Through the mid-1930s they tested numerous elements and defined various nuclear reactions. They also discovered that if the velocity of neutrons was reduced through collisions with water or some other substance that contained hydrogen, the slow neutrons were more effective than fast ones in initiating the reactions. The interaction of neutrons with uranium was particularly puzzling. A number of different products seemed to be formed, as detected by different half-lives of radioactivity, yet their identity was unclear. Behind the mystifying behavior of uranium was the recognition that, if neutrons adhered to this heaviest of the elements that naturally occur on earth, and the even heavier new nucleus then emitted a beta particle, the result would be an artificial element, one unit higher in the periodic table than uranium at 92. Since several betas were detected, it seemed possible that more than one new element was being produced. [97]

Besides Fermi, Irène Joliot-Curie, the daughter of Marie and Pierre Curie, tried from Paris to untangle this confusion of products, as did Otto Hahn's circle in Berlin. Hahn was the world's most experienced radiochemist, having begun his discovery of new radioelements in the first decade of the century. His important colleagues were the physicist Lise Meitner and the chemist Fritz Strassmann. But before we get to their discovery of nuclear fission, we must trace the story of the concept that there was usable energy locked in the atom.

Atomic Energy

Ernest Rutherford made newspaper headlines in 1933, not for anything he did but for his assertion that something could not be done. He had been irritated by a lot of wild speculation, leading him to tell a meeting of the British Association for the Advancement of Science that "any one who says that with the means at present at our disposal and with our present knowledge we can utilize atomic energy is talking moonshine." [18:197] Superb physicist that he was, he was correct. But some have ignored the qualifications Rutherford voiced and have claimed this as one of the very few times in his career when he was wrong. The qualifications are important, however, because knowledge and apparatus in 1933 for extracting the atom's energy, while far better than at the turn of the century, still were inadequate.

As mentioned above, the ability of radium to tint glassware, burn flesh, and glow in the dark signified an unusual outpouring of energy from radioactive substances, and measurements showing its temperature higher than its surroundings quantified the amount. An American chemist, H. C. Bolton, in 1900 wondered:

Are our bicycles to be lighted with disks of radium in tiny lanterns? Are these substances to become the cheapest form of light for certain purposes? Are we about to realize the chimerical dream of the alchemists,—lamps giving light perpetually without consumption of oil? [18:198]

Rutherford's colleague, Frederick Soddy, speculated in 1902 that radioactivity might be the source of the sun's heat (ordinary combustion was inadequate). The next year Soddy wrote that the earth was a "storehouse stuffed with explosives, inconceivably more powerful than any we know of, and possibly only awaiting a suitable detonator to cause the earth to revert to chaos," thus calling attention to the potential dangers of the new science. [18:198] This was novel, for scientists had never had occasion to fear such a catastrophic outcome from their investigations. Fueled by other concerns, over the next decades the idea of science and social responsibility slowly grew. Most scientists took the comfortable position that nothing should suppress research; this was the "pure" search for knowledge. But applications, which were normally out of their hands, should be controlled if dangerous. [18]

Soddy's disquietude was largely dismissed by Rutherford, whose mind was less conjectural. He made the mistake, however, of joking with a colleague that, "could a proper detonator be discovered, an explosive wave of atomic disintegration might be started through all matter." [18:198] The colleague, who seems to have taken him seriously, put it into print. In this way, Rutherford's name was linked with the fear that "some fool in a laboratory might blow up the universe unawares." [18:198] (And the image of the mad scientist in science fiction novels, films, and comic books gained a hook into reality. [110])

Soddy's insight was timely, for when the amount of energy in one gram of radium was journalistically transformed into terms the layman could comprehend, it was far more impressive than the tiny needs of a bicycle lantern. That gram could lift 500 tons a mile into the air, or power a 50-horsepower car at 30 miles an hour around the globe. Yet, while radioactivity found application in medicine, in paint that made objects glow in the dark, and in explaining the geological age of the earth, no one attempted to harness the atom's energy for power. Their goal was to understand the laws of nature, even if this did not hinder their speculation. [9; 16; 18]

Soddy continued to lecture and write on this issue, stressing that radioactive transformations occurred with explosive violence and that the store of energy was far larger within atoms than in chemical reactions. A notable distinction, however, was that the decay of one radium atom had no effect whatever upon another radium atom nearby, while the detonation of a molecule of chemical explosive did cause nearby molecules to explode. Rutherford, serious and conjectural this time, in his classic text *Radio-Activity* (1904), reasoned that the atoms of all elements, not just the radioactive ones, should possess enormous stores of energy. [18]

British chemist William Ramsay echoed this idea. The supply of radium, he said, was small, but if a catalyst was found to make all elements transform at a rapid rate, "the whole future of our race would be altered." [18:199] Pierre Curie, in his Nobel lecture (delivered in 1905, but for the 1903 prize), with radium's ability to burn healthy, as well as diseased, tissue on his mind, warned that "radium could become very dangerous

[handwritten: scientists are conflicted → diff. opinions → don't take science as one entity]

[handwritten: widespread infiltration of science into society ↳ disruption b/w scientists cannot into society]

in criminal hands." [18:199] Novelist H. G. Wells picked up these threads in his book *The World Set Free* (1914), which described powerful atomic bombs. Soddy and others, in the gloom of the World War I period, stressed that the energy that now "oozes" from radioactive bodies is on the order of a million times greater than that in an equal weight of chemical explosive, and people must learn to avoid war before they discover how to speed that process. As strong as Soddy's belief that atomic energy would be harnessed was Rutherford's conviction that it would not; to an old friend he confided: "If it had been feasible it should have happened long ago on this ancient planet. I sleep quite soundly at nights." [18:200]

By the early 1920s, it was possible to hang more scientific detail on these vague pronouncements. Einstein's $E = mc^2$ relationship was combined with precision measurements of atomic weights. In numerous nuclear reactions the mass of the products was found to be less than that of the initial substances. The lost mass had been converted to huge amounts of energy (huge, that is, per atom; only a few atoms at a time were involved here). British astrophysicist A. S. Eddington suggested that the vast energy of the sun and other stars derives from one such reaction: the fusion of four hydrogen atoms into one of helium. He added his concern about whether such energy, if harnessed on earth, would be used "for the well-being of the human race—or for its suicide." [18:200]

The British physicist F. W. Aston, upon whose meticulous atomic weight measurements this new wave of concern was based, was more sanguine. He felt that the release of atomic energy might well be the greatest achievement of the race, providing it with unlimited energy. The largest ocean liner, for example, could be propelled across the Atlantic and back at full speed if the hydrogen in a pint of water could be transmuted into helium. Aston, further, condemned the suggestion that research in this area be suppressed because of its potential danger; one could not stand in the way of progress. He was supported in this defense of research by British biochemist J. B. S. Haldane, whose real purpose was to urge the investigation of relatively more humane gas warfare, then suffering from disrepute in the wake of World War I. Haldane cloaked his urgings somewhat under the mantle of the social responsibility of scientists. Much more specific about social responsibility was the Russian mineralogist Vladimir Vernadsky, who in 1922 argued both that control of atomic energy was inevitable and that "scientists must not close their eyes to the possible consequences of their . . . work." [18:201]

From outside the world of science, the bishop of Ripon declaimed in 1927 that a sense of social responsibility in the laboratory was not enough. The laboratory should be closed! "The world is going too fast and . . . humanity would be benefitted if physicists and chemists suspended operations for ten years," using this moratorium to improve human relations. [18:202] The bishop was not a voice in the wilderness but a spokesman for a viewpoint that had long existed: that there was an arrogance in the acquisition of scientific knowledge that usurped people's responsibilities to society. This was the perspective seen by some in the story of Galileo's struggle with the Inquisition (he should have moderated his ideas so as not to threaten the church), in the stories of Faust and Dr. Frankestein, in the development of poison gases in World War I, and most recently in the trial of Thomas Scopes for teaching evolutionary biology in a Tennessee school. [18]

No one reacted with more hostility to the bishop's remarks than the head of the California Institute of Technology, Robert A. Millikan. The Nobel laureate's anger, however, was directed more at fellow laureate Soddy, who was primarily responsible for raising the "hobgoblin of dangerous quantities of available subatomic energies." [18:202] Millikan was convinced "that the creator has put some foolproof elements into his handiwork and that man is powerless to do any titanic physical damage." [18:202]

The advent of particle accelerators, and especially Cockcroft and Walton's success in "smashing the atom," gave hope that the ability to harness the energy was near. That time had not yet arrived, though, for it was widely recognized that far more energy was required to run the accelerators than was released in the relatively few nuclear reactions produced. In the early 1930s, the newly discovered neutron offered another tool. Russian theoretical physicist Lev Landau regarded useful nuclear energy from reactions with charged particles to be science fiction. But this could change to reality if a reaction were discovered in which a neutron released other neutrons in an ongoing chain reaction. [18]

In such a conjectured process, the neutron would not have to be accelerated artificially; indeed, it might even be useful to slow it down. The neutron would split a nucleus, releasing not only energy but other neutrons as well. If each of these newly produced neutrons then struck other nuclei, causing further disintegrations and still more neutrons, an expanding series would occur. This process, later called a chain reaction, may be likened to the structure of a tree, which grows from a single trunk to several major limbs to many more branches. However, unlike a tree, whose parts are of different size, each nuclear "junction" would release about the same amount of energy. If this occurred quickly, the energy would be liberated explosively.

Einstein joined the debate in 1934, stating that it was fruitless to search for atomic energy. At the Nobel Prize ceremonies the next year, physicist Frédéric Joliot-Curie (Irène's husband) was more hopeful. Eddington doubted success; Prometheus's precedent of stealing fire from the gods for mankind was not likely to be repeated. But, mindful of international tensions in Europe, he admitted that the tiny cloud on the horizon was ominous. Niels Bohr, the distinguished Danish physicist, saw no reason for concern. He felt that harnessing the atom's energy was becoming less and less likely. Even if a nucleus could be shattered, this would not mean that one had a weapon, much less a controlled source of energy. Rutherford, despite his "moonshine" comment, managed to convey a nagging feeling to the secretary of the Committee of Imperial Defence that nuclear transformations might someday be important to Great Britain's security. Rutherford, impressed with Fermi's work in Rome, also wondered if the neutron might be the "magic bullet" that would allow a profitable smashing of atoms. [18]

The point of this recital of views about useful atomic energy is to show that the discussion persisted rather vigorously for four decades prior to the discovery of nuclear fission, which then precisely showed the way. It was not a sideshow of a freakish topic but one taken quite seriously by the scientists involved and by the public. It could not be otherwise, considering the scientists' collective distinction: Nobel laureates, present and future, included Rutherford, Curie, Ramsay, Soddy, Aston, Fermi, Millikan, I. I. Rabi, Lawrence, Einstein, Landau, Joliot-Curie, and Bohr. [18]

With hindsight, we see that the discussion, while serious, was not especially innovative or profound. Recent discoveries were incorporated into the arguments, but some ideas were strangely absent. Landau touched upon it, but only Leo Szilard really explored the concept of a chain reaction, and he kept it secret. No one seems to have speculated about the splitting of atoms having the largest masses, such as uranium and thorium, though the technical data were available to show that fission, as well as fusion, would convert mass into energy.

Scientists and the public, it seems, lacked a high degree of fanciful thinking. At the same time, many showed an almost blind confidence in the future conquest of nature, nurtured no doubt by three centuries of Baconian success. Even during the depression of the 1930s, most people had confidence that scientific knowledge was bound to be beneficial in the long run. While scientists could more accurately than others predict some future events in their own fields, it was not a pastime in which they often engaged. Scientists, it must be understood, even the highly creative ones, usually behave as ordinary human beings, and the human condition is basically one of hope, with few glimpses of distant peaks or even the paths toward them.

Nuclear Fission

As mentioned above, when uranium was bombarded by neutrons some radioactive substances were created. Several different beta particles were emitted from them, which meant that the new nuclei were left increasingly more positively charged. Since nuclear charge determines the identity of an element, this suggested that elements heavier than uranium were being created. It was to determine the properties of such products that radiochemists such as Irène Joliot-Curie and Otto Hahn became involved. A standard technique, dating back to the discovery of radium in 1898, was to use a "carrier" element. When the number of atoms of the unidentified radioelement were too few to permit chemical identification by normal laboratory methods, it would be mixed with other inert elements to give enough material to see and handle. Various dissolutions and precipitations would be performed on the mixture, constantly tracking where the unknown went by detecting its radioactivity.

Before long, it would become apparent that a known element was being concentrated, and the unknown was moving along with it, being chemically similar. The known element was called the carrier (barium worked well with radium), and purifying it would also concentrate the unknown. Indeed, once a good carrier was found, more of it would be added to the mixture, for ease of separation. Eventually, another technique could be used to isolate the new substance from the carrier (fractional crystallization to separate radium from barium). [49]

In the late 1930s, radiochemists working on uranium faced the problems of a confusing number of decays and an inability to separate some unknowns from carriers. Adding to the difficulties, there was more than one kind of radium and of uranium. The varieties are called isotopes. Uranium's best-known isotopes are U-235 and U-238. The number gives the atomic mass of the isotope, composed of the sum of all its protons and neutrons. Since all types of uranium have 92 protons (remember that the electrical charge determines the element, and each element has distinct chemical properties), U-235 has 143 neutrons in its nucleus (= 235 − 92), and U-238 has 146 neutrons (= 238 − 92).

Scientists do not enter their laboratories with blank minds. While they should be open to the significance of unexpected results, they must begin their experiments with some concept of the process under examination and enough of a guess of its magnitude to prepare the proper instruments to record it. These radiochemists operated under the hypothesis that beta decay from uranium produced new elements that were heavier than uranium: transuranics. The alleged transuranics were themselves radioactive, also emitting beta particles. A most serious challenge to their working concept, however, was that some of the decays seemed to be emitted by radium isotopes. Radium is not far from uranium in the periodic table of elements, but it is lighter than uranium.

European politics forced its way into this scientific riddle. Hitler's laws against Jews interrupted the decades-long collaboration of chemist Otto Hahn and physicist Lise Meitner and forced her to flee Berlin for Sweden. Hahn and Fritz Strassmann continued the investigation of the curious radium isotope, and in December 1938 they were amazed to realize that its activity was separating along with the barium carrier, not the radium. Hahn's long experience with this "classic" separation process gave him confidence that no mistake had been made. Somehow they had produced not radium but a new kind of radioactive barium, roughly half radium's weight. They speculated that, if uranium split in two, elements with weights in the middle of the periodic table, such as barium, would be formed. But because no one had ever seen a reaction in which anything heavier than an alpha particle was emitted, they recognized that this was a "drastic step which goes against all previous experience in nuclear physics." [18:206; 49]

Before the paper they composed was published in January 1939, Hahn wrote to Meitner with the news. She shared the letter with her physicist nephew, Otto Frisch, visiting from Bohr's Copenhagen laboratory. Together they reasoned that the neutron might give enough energy to a uranium nucleus to deform it into a dumbbell shape, and it might then split at the narrow neck. A back-of-the-envelope calculation showed that the uranium and neutron together had more mass than any combination of two elements from the middle of the periodic table. The lost mass appeared as energy, and its amount could be calculated by Einstein's formula that related mass and energy. In this case the enormous amount of 200 million electron volts (MeV) of energy would be released. The significance of this can be comprehended when it is understood that only a few electron volts are released in the most energetic chemical reactions. Nuclear reactions thus involve millions of times more energy. [18]

The Hahn-Strassmann revelation was made by chemical means. Frisch returned to Copenhagen and used physical techniques to confirm it.

With apparatus to observe the energy released in these nuclear events, he quickly detected large bursts of energy coming from neutron-irradiated uranium. A biological colleague told him that the process of cell division was called fission, and this term now was applied to the atomic nucleus. Early in the new year, Frisch wired the news to Bohr, who was on a trip to America, and it was shared with a small group that included Enrico Fermi. The Italian physicist had just used the occasion of the award of the Nobel Prize to escape his homeland, for its fascist dictator, Mussolini, was giving signs of racial persecution, and Fermi's wife was Jewish. Fermi now held a position at Columbia University. Before January 1939 ended, fission was confirmed in at least four American laboratories and by Frédéric Joliot-Curie's team in Paris. It was the most electrifying scientific discovery in years. [18; 39; 97]

But why was fission so breathtaking? Had not scientists, and novelists too, discussed the extraction of energy from the atom for several decades? The answer is multifold, part scientific and part political. New phenomena of nature are not discovered every day, and fission appeared to be one of great significance, worthy of the rapt attention of the scientific community. Moreover, earlier efforts to detect chunks of matter, larger than alphas, that broke off from atoms had failed, so it was now unexpected. With fission, however, the "chunk" could be said to be each half of the uranium nucleus. Measurement of the masses of such middleweight elements, compared with the combined mass of uranium and a neutron, showed that large amounts of mass would be lost in the reaction, converted to energy. Fission, therefore, suddenly made usable nuclear energy seem a long step closer. But political turmoil in Europe also was closer. War clouds gathered as Hitler built up his military machine in the 1930s, and fear that the still powerful German scientific community might provide him with weapons that used nuclear energy was prominent in the minds of scientists who had fled Germany. [18]

Something else was needed, though, for this energy to be harnessed: a chain reaction. This was a concept familiar to chemists, who recognized that a firecracker or molecule of explosive triggered its neighbors, and these in turn ignited more neighbors, until the supply was consumed or the pieces blown far enough apart that they were no longer neighbors. Physicists had less occasion to deal with such a process. Rutherford, for example, had pointed out that nuclei are so far apart, even in solid matter, that the radioactive decay of one has no effect on its neighbors. In particle accelerators, too, the nuclear disintegrations are produced by a relatively inefficient beam of projectiles. None of this gave physicists reason to be hopeful about using nuclear energy. [18]

The "magic bullet," the neutron, changed this picture. If neutrons were emitted as nuclei fissioned, and if enough of these neutrons struck other nuclei to keep the reaction going at a steady rate, generation of neutrons after generation, a power source could be made. If the reaction proceeded at an expanding rate (more and more nuclei split in each succeeding generation), there would be an uncontrolled release of energy—an explosion.

When Frisch explained the concept of fission (not the chain reaction) to Niels Bohr, the great Danish physicist struck his forehead and exclaimed, "Oh, what fools we have been! We ought to have seen that before." [18:208] That was the reaction of most physicists. There was an exception, however: Leo Szilard. One of the most creative and prescient minds of the century, this Hungarian refugee from Nazism conceived in 1934 (almost five years before the discovery of fission) of a process in which an element absorbed one neutron and emitted two: a chain reaction. Although he did not know what element would be the key, and he did not specifically predict the fission process, he was sufficiently concerned with this "formula" for a gigantic bomb that he took out a secret patent on it in Great Britain and assigned it to the Admiralty, thereby hoping to prevent publication of the idea. When uranium fission was discovered, the pieces of Szilard's puzzle fell into place. By this time Szilard was in New York, in contact with Fermi and others studying neutron reactions at Columbia. [18]

During the next few months neutron investigations went in many directions in various laboratories. Neutrons slowed down by water, or by some other "moderator" such as carbon, were found to be more effective in fissioning uranium than were fast neutrons. Quantitative measurements were made of the energy released. Tentative confirmation was obtained that two to four neutrons were emitted in each fission (the correct value is about 2.5). Uranium was seen to fission in a variety of ways, as numerous "fission fragments" were detected, roughly but not exactly half the weight of the uranium atom. Bohr suggested that the rare isotope U-235, present in natural uranium to the extent of only 0.7 percent, was fissioning, and not the far more abundant U-238 (99.3 percent of uranium in minerals). Investigations, thus, were begun on methods to separate these two isotopes. This was a difficult problem because chemical processes were ineffective, since isotopes are chemically identical. A variety of physical techniques was tried. In an effort to create a chain reaction, Fermi and Szilard began to plan construction of "piles": structures of carbon moderator in the form of graphite bricks, interspersed with pieces of uranium. The carbon would slow neutrons and optimize their collision with the uranium. [18]

For most physicists the excitement of basic scientific discovery was accompanied by realization of the potential uses of nuclear energy: reactors and bombs. A reporter from the *Evening Star* attended a physics conference in Washington, DC, at the end of January 1939, and the revelation of fission by Bohr and Fermi led to page-one headlines: "Power of New Atomic Blast Greatest Achieved on Earth" and "Physicists Here Hail Discovery Greatest since Radium." Other newspapers and newsmagazines picked up the story and maintained a continuing interest in the subject, always ready to emphasize the potential of nuclear explosives. The *New York Times* science editor, for example, speculated that a Martian observing earth's cataclysmic end might comment, "Some imbecile has been annihilating matter." [18:214]

Manhattan Projects

Early Steps toward a Bomb Project

SOME SCIENTISTS FELT that the U.S. government should be alerted to these new developments, so in March 1939 Fermi visited Washington. Navy officers and civilian scientists from the Naval Research Laboratory were interested primarily in nuclear energy as a power source for the propulsion of submarines; since oxygen was not required in the reaction, submarines could remain submerged indefinitely. Consequently, the navy offered $1,500 to support nuclear research, not an inconsiderable grant in those days. But scientists were much more concerned with basic research and explosives than with propulsion. Anyway, much had to be accomplished before a controlled power source could be built.

Quite early in 1939, it occurred to Szilard that self-censorship was a good idea. Why give the Nazis valuable information? Others were doubtful. At that point a chain reaction seemed so far away that such an assault upon the cherished tradition of openness in science was unacceptable. Szilard was persistent, nevertheless, and more convincing by spring, when neutron emission in each fission—a necessary condition for a chain reaction—was confirmed. Voluntary censorship was agreed to by many physicists in several countries, but their pact unraveled when Joliot-Curie's team ignored it. The desire for priority proved stronger at that time, the fame from making a notable discovery being the traditional "coin of the realm" in the scientific community. Not until September 1939, when World War II began, did the belligerent nations impose secrecy, and not until mid-1940 was self-censorship effective in the United States. [18]

There were valid reasons for concern about developments in Germany. Fission, of course, was discovered there, and the Nazi-inspired "brain drain" had not emptied the country of all its highly talented scientists.

27

In June 1939 a paper appeared in a prominent German periodical entitled "Can nuclear energy be utilized for practical purposes?" To the author the answer was yes, and it was obvious that "the energy liberation should . . . assume the form of an exceedingly violent explosion." [18:215] Indeed, more than a month before this, one team of German scientists contacted the Reich Ministry of Education and another the War Office, both alerting the bureaucracy to fission's potential. The government responded with haste, and Germany alone had a military office solely concerned with nuclear energy at the outset of World War II. [107]

Not that other nations were idle. The news of neutron emission in the fission process that had inspired the Germans also energized Soviet scientists. If a chain reaction was theoretically possible, it was obvious that explosives also were plausible. Though interrupted by the German invasion of their country, by the time of Hiroshima and Nagasaki in 1945, Soviet scientists were well along in their own nuclear weapon project. The British were even faster off the mark. With unusually good links between scientists and government, a research program soon was underway, although once war broke out in September 1939, it received lower priority than weapon systems that promised a more immediate return, such as radar. [45; 46]

By early summer 1939, Szilard was frustrated that neutron research in the United States was not coordinated and being pursued more aggressively. Fermi, for example, left New York for a few months to study not nuclear fission but cosmic rays. With fellow Hungarian Eugene Wigner, a physicist at Princeton University, Szilard decided that some political step must be taken. When an acquaintance with ties to the White House offered to deliver a message to Roosevelt, they seized the opportunity. To catch the president's eye, they enlisted the aid of the world's most famous scientist, someone the president could not ignore. Thus, in August, Szilard and Albert Einstein composed a now famous letter to Roosevelt, signed only by Einstein. [52:16–17; 111]

The situation was novel, not only because refugees from Nazism directly contacted the leader of the country that gave them sanctuary but because academic scientists and government had so little experience with each other. Scientists had always been protective of their intellectual independence and could look apprehensively at Germany, where the government did take an interest in biology (the master race concept) and in "Jewish physics." The federal government, for its part, had for a century and a half been unable to make up its mind about the value and constitutional propriety of underwriting research in the universities. Einstein's letter, therefore, described the fear that Hitler might acquire

uneasy relationship blw science/govt
only want support → first step in developing relationship

nuclear weapons but contained the barest of requests. No financial support, bureaucracy, or government program was solicited, merely a liaison with the physicists to keep the administration informed. Federal blessing, not money, was sought. [18]

The letter was not delivered until October, the courier waiting to find a lull in the president's activities following the outbreak of war in Europe. Roosevelt appears to have comprehended immediately the significance of the message; he referred the matter to a newly formed Advisory Committee on Uranium. At its meeting ten days later, several academic physicists described the state of neutron research to a few military ordnance experts. Almost incidentally, it seems, talk turned to the question of government financing. An army officer lectured the scientists that troop morale, not new weapons, won wars, to which Wigner gently responded that the army should therefore not object to a cut in its procurement budget. "All right, you'll get your money," the officer growled. Szilard thereupon ceased his efforts to secure funding from large industrial corporations and private philanthropists, and in this casual fashion the government eased into the role of paymaster on a gargantuan scale for scientific research. However, if the Einstein letter had never existed, events would probably have flowed in pretty much the same order. History is determined by social and cultural forces as much as by individuals, and it was inevitable that the potential of nuclear weapons would have been brought before the government. [52:20]

Through all of 1939, information about fission and the possibility of an explosive chain reaction appeared widely in scientific publications and the popular press. Some 100 scientific papers were printed during the year, which was crowned by the publication of two major surveys, in the *Reviews of Modern Physics* and for the Chemical Society of London. While the Allied effort to build nuclear weapons, what was later called the Manhattan Project, would be cloaked in secrecy, it is clear that the early research and discussion of these enormous bombs was anything but private. For in the political climate of that year, they were regarded as remote, and therefore easily forgotten. To those who continued their investigations in this field, moral or ethical questions rarely surfaced. They did not worry about whether they personally should work on weapons or whether they should turn their beloved basic science toward such an application. They spoke of "if" a bomb could be made, not "when." There were so many scientific hurdles to overcome that they believed the next generation would face the hard questions, not theirs. In any case, the only moral question they perceived was whether they could allow Hitler to acquire nuclear weapons.

Coordination of Research

One scientific hurdle, overcome early in 1940, was proof that U-235 was the species of uranium, the isotope comprising only 0.7 percent of natural uranium, that slow neutrons caused to fission. If enough of this rare material could be separated, it might power a controlled chain reaction. Fermi, who felt that natural uranium (mostly U-238) was still a possible fuel, was provided 4 tons of pure graphite and 50 tons of uranium oxide by the Advisory Committee on Uranium for chain reaction studies at Columbia University. For a bomb, however, it was recognized that fast neutrons were required, or the uranium metal would blow itself apart before much of an explosion occurred. It was not known if U-235 would fission with fast neutrons.

The relatively unhurried pace of investigation was given a jolt in the summer of 1940, when Vannevar Bush, an engineer who was president of the Carnegie Institution of Washington, a private research laboratory, persuaded President Roosevelt to create the National Defense Research Committee (NDRC) and appoint him its director. Anticipating America's involvement in the war, Bush wanted to marshal the nation's scientific talent for work on weapons. Uranium-related studies were but one project under the NDRC, though it became the largest. A measure of the seriousness and energy now injected into nuclear research is the veil of secrecy that descended and the availability of significant government funding: the agency opened for business with $100,000. A year later, when Bush sought to make the large bureaucracy he had created more efficient, Roosevelt named him head of the new Office of Scientific Research and Development (OSRD), part of the Executive Office of the President and with greater access to the White House. The president of Harvard University, chemist James B. Conant, became head of the NDRC, now under the OSRD. With government assuming a greater role in nuclear research, the refugee scientists were eased from positions where they had a broad view of activities and a voice in policy matters. They were, after all, enemy aliens (once the United States went to war), and information was increasingly held to a need-to-know basis. [52:24–25, 51]

Isotope separation methods were given much attention in 1940, as were studies of the amount of material needed (the critical mass) to sustain a chain reaction. Fermi and Szilard had difficulty obtaining satisfactorily pure uranium metal (the impurities absorbed neutrons, removing them from chain reactions) and tried purer uranium oxide instead. Graphite moderator, to slow the neutrons, also contained too many impurities, and commercial sources were pressed to improve the quality. This industry, like many others, had little prior need for great purity in its products (certainly not for the "lead" in pencils) and had to develop new techniques to meet the new "nuclear standards." As these improvements were made, Fermi and Szilard built a series of "piles," the early term for nuclear reactors. As purity of the materials improved, the researchers approached a chain reaction closer and closer and learned the necessary size and shape of the reactor. [52:54–56]

While this effort with natural uranium was underway at Columbia, other scientists, there and at other universities, studied means to isolate the U-235 isotope. The reasoning was that the bulk of the chain-reacting weapon or reactor would be smaller with the fissionable isotope concentrated, and the process would be more efficient. Isotope separation, however, is very difficult; as mentioned above, chemical techniques are of no avail, and laborious physical methods (most barely tried before) must be used. The most familiar concept was that used in a centrifuge, and long tubes filled with a gaseous compound of uranium were spun rapidly to separate the light and heavy components. But since the separation in each tube was minute, the gas slightly enriched in U-235 would have to be pumped to the next centrifuge, and so on in a long chain, becoming marginally richer in each stage. This process faced severe structural problems, however, as the more efficient long tubes were insufficiently strong, and they also experienced destructive vibrations at certain frequencies as they spun up to their operating rotational velocity. [52:63–65, 96–97]

An alternative called gaseous diffusion was similar in requiring a long chain of stages in which the product would be increasingly enriched. The feed material would be forced under slight pressure into a stage, and the lighter gas molecules would have a greater likelihood of passing through a screen that contained microscopic holes. This process, too, had herculean difficulties. Uranium forms only one gaseous compound, uranium hexafluoride, which is highly corrosive and toxic. All interior surfaces—stages, pipes, pumps, and the porous barrier—had to react less to this gas than stainless steel does to air, or the corrosion products would clog the barrier. Nickel was the material of choice, but it took much time to learn how to construct the components. [52:97–101]

Liquid thermal diffusion was another technique to separate isotopes; here the lighter and heavier molecules would migrate to hotter or colder regions of a heated tube. Again, this would require many stages. The one method that was not a statistical process was electromagnetic separation, in which, theoretically at least, total division of the isotopes should be possible in a single operation. Moving, charged particles are bent into circular paths in a magnetic field, and the curvature of the path depends upon the particles mass. So each isotope should follow a distinct path, according to its mass. This is the principle of the mass spectrometer, a familiar tool in physics laboratories for a few decades, and of the cyclotron, invented by Ernest Lawrence, of the University of California at Berkeley. Ideally, two buckets could be placed in the evacuated chamber, one to catch U-235 ions and the other U-238. In practice, it was far from being so simple. All the particles moving in their curved beams carried an electrical charge, which made the beams fuzzy as the particles repelled each other. Further, the uranium ions tended to settle on the surfaces all over the chamber's interior, and a huge and difficult chemical operation was required to reclaim the precious product. [52:56–60, 162–67, 294–98]

Because experience with these processes was of laboratory scale at best and theoretical at worst, success was uncertain with any. Hence, each was pursued until it might be proven impossible or another proven satisfactory. An attack on such a broad front was unusual in peacetime industry or a government program, where cost was a controlling factor, but it was characteristic of the bomb project and of other wartime efforts, such as the production of synthetic rubber.

By the summer of 1941, none of these problems was solved, but at least they had been identified and a program established under the OSRD to coordinate their investigation. The American effort, which experienced cycles of enthusiasm and pessimism, was buoyed by reports from England in the spring of 1940 (by Frisch and a colleague, Rudolf Peierls) and a year later (the so-called MAUD Committee) which suggested that the critical mass of U-235 for a weapon was smaller than believed earlier. [46; 84] This meant that enough material could likely be separated within the anticipated duration of the war, and the bomb would be small enough to transport it. In addition, there was better evidence that U-235 would fission with fast neutrons, necessary for an explosion. [52:32, 42]

Most interesting was the discovery in 1940, by Edwin McMillan and Philip Abelson, using Lawrence's Berkeley cyclotron, of the first transuranic (an element beyond uranium in the periodic table), which formed when U-238 captured a neutron. This man-made element, with 93 protons in the nucleus, was called neptunium, after the planet Neptune, just as uranium had been named after Uranus. Soon, a Berkeley team led by nuclear chemist Glenn Seaborg discovered the next higher element, number 94, exhausting the solar system's outer planets with the name of plutonium. Theory predicted that plutonium should fission even more readily than U-235, and since it was a distinct element, it should be separable by chemical methods. This meant that they need not rely solely on U-235 for a weapon. Yet the idea of constructing a bomb from a material that does not exist normally on earth had to be one of humanity's most remarkable gambles. [52:33–34]

Progress was sufficient on many fronts by autumn 1941 for Bush to meet with Roosevelt and Vice-President Henry Wallace. Roosevelt, recognizing that the work was still in an exploratory stage, authorized expanded research and planning but not construction. Bush was told to preserve secrecy and to restrict policy discussions to Roosevelt, Wallace, NDRC head Conant, Secretary of War Henry Stimson, and Army Chief of Staff General George Marshall. [52:44–46]

Bush began to draw up plans for the large factories necessary to separate isotopes of uranium. Fortunately, America, with a strong tradition of chemical engineering going back to World War I, had much experience in designing technically complicated plants. The director of the Bureau of the Budget assured Bush that money would always be available from the president's discretionary funds, if the OSRD budget was inadequate. In time, some members of Congress were informed of the nuclear program, so they could persuade their colleagues to vote appropriations for the president's "deep pockets." One may wonder how cooperative these officials would have been had they been able to forecast that the atomic bomb would cost over $2 billion. [52:52]

The Met Lab

Japan attacked the U.S. naval base at Pearl Harbor, on the Hawaiian island of Oahu, on December 7, 1941, "a day that will live in infamy," as Roosevelt said when asking Congress to declare war on the Axis powers. Just one day before the raid the president authorized a significant expansion of the nuclear program. Ernest Lawrence was given $400,000 for large-scale studies at Berkeley of electromagnetic separation, $500,000 was allocated for reactor materials, and other work at Columbia and Princeton received $300,000. This was a far cry from the initial grant of $6,000 that the Einstein letter had squeezed from the government, and a quantum jump even from the budget of a few months earlier. [52:52–54]

Partially to protect the nuclear efforts by moving them away from the nation's coastline, but more to achieve the efficiency of consolidation, most activities relating to plutonium and to the bomb were relocated on the campus of the University of Chicago. Nobel laureate Arthur Compton was placed in charge by Bush and Conant, and the administrative structure was called the Metallurgical Laboratory, or Met Lab. The name was considered sufficiently uninteresting that strangers would not be curious. (The Germans discouraged inquiries at their project by putting up a sign saying VIRUS HOUSE.) Compton quickly set his goals: determine if a chain reaction was possible, by July 1942; achieve that chain reaction, by January 1943; extract the first plutonium from the uranium in which it "grew," by January 1944; construct a bomb, by January 1945. [52:53–56]

Encouraging results were reported early in 1942. The critical mass of a sphere of U-235 was estimated to be between 2.5 and 5 kilograms (5.5–11 lbs.), a welcome reduction from the previous range of 2 to 100 kilograms (4–220 lbs.), and still more so from one of the first approximations of tens of tons (of U-238). Just as important as the small size was the growing conviction that the material could be detonated efficiently, yielding as much energy as the equivalent of 2,000 tons (or 2 kilotons) of TNT. Much depended upon the ability to keep the fissionable material together long enough for many generations of fissions to occur, until the multiplying fast neutrons could reach many atoms. The small size made this achievement more likely. [52:61; 88:321]

As the summer of 1942 approached, Bush informed Roosevelt that pilot plants should be built to test several production processes. The project was changing from one of scientific research and development to exploration of the technological production effort, with a weapon as the goal. It was clear that the scientists lacked the knowledge and experience to manage such large-scale activities, much as they were supremely (and perhaps arrogantly) confident in their own abilities. The Army Corps of Engineers was the logical agency to supervise construction of the factories, roads, and cities on the drawing board. But the army was not very enthusiastic about taking on this job, for it was otherwise occupied, and it was reluctant to give the necessary high priority for personnel and materiel to a project whose outcome was uncertain. Its ambivalence was reflected in the months it put off acquiring a site in Tennessee that Bush had recommended for a gaseous diffusion plant. [52:67–75]

The situation changed markedly in September 1942, when Leslie Groves, a newly appointed brigadier general, with vast construction experience (including building the Pentagon), took charge. A large and corpulent man, with an abrasive style that irritated the scientists and led to frequent clashes with them, he was nonetheless well suited for the job. (Despite his girth, he has been portrayed in two feature films by Hollywood idols Brian Donlevy and Paul Newman.) The scientists disliked not only Groves, the man, but the organization he represented. They had successfully resisted efforts to put them into uniform, which was the price that scientists in World War I had to pay to work for their country, and now feared that the nonintellectual "military mentality" and the imposition of tighter security regulations would hinder their efforts. Groves, however, had the self-confidence to overcome such personnel problems and, indeed, the courage to put his career at risk by making decisions involving processes and apparatus before they were developed and shown to work. He immediately got for himself the authority to order scarce materials with the highest priority and purchased the site at Oak Ridge, Tennessee. Because the army's first contact with nuclear work was at Columbia University, through its office in Manhattan, the entire effort now was given the code name of that unit, the Manhattan Engineer District (popularly called the Manhattan Project). [52:81–83]

With the commitment to a pilot-plant building program, contracts were signed with Stone and Webster, a large engineering construction company, and with Du Pont, the chemical giant known for its skill in designing apparatus and plants. Du Pont, with no experience in nuclear physics, was reluctant at first, but Groves was persuasive. Aware that the Krupp armaments empire in Germany was called a "merchant of death" after World War I, Du Pont sought to avoid a similar public relations disaster by insisting on a profit of only one dollar. Soon Du Pont chemical engineers were learning about chain reactions and pondering the methods of producing fissionable material in quantity. When the bomb project is discussed, the focus is usually upon the scientists' work. It is worthwhile, however, to reflect that the role of industry was not only necessary but equally impressive in overcoming hurdles. [48; 52:91, 187]

One particular obstacle faced by the scientists was the chain reaction. Without proof that this could occur, there was little sense in proceeding with all the other plans. Fermi, by now at the Met Lab, continued to build piles of graphite bricks and natural uranium. As the materials were obtained in greater purity, as the reactor's geometry improved, and as the size increased, he approached closer and closer to a sustained chain reaction, with at least one neutron from each fission causing another fission. The next structure incorporating these modifications, Fermi calculated in the autumn of 1942, would achieve a chain reaction. Compton planned that this pile be built in the Argonne Forest Preserve, near Chicago, in case the reaction got out of control and spewed lots of dangerous radioactivity (such reactors are incapable of exploding as a weapon does). But labor difficulties had delayed construction of the building at Argonne, and Compton accepted Fermi's assurances that his control rods of cadmium, and other techniques to soak up neutrons, would safely harness the pile. Chicago Pile–1 (CP–1), the first reactor to go "critical," would be built under the stands of Stagg Field, the university's football stadium, in the middle of the city. [52:108–12]

(Today commercial power reactors are sited as far from urban areas as possible, but some low-power experimental reactors still operate— not without protest—on university campuses in cities. During wartime, however, it is more common to have decisions made that bear potentially dangerous consequences, and of course the public is usually ignorant of them.)

Through November 1942, Fermi and his crew of young scientists pushed further into the realm of Big Science. Layer after layer of graphite bricks were emplaced in a structure that looked a bit like a 20-foot cube with its corners removed and supported by wood bracing. Higher-quality uranium was placed in a pattern toward the center. In all, the pile consumed 400 tons of graphite, 6 tons of uranium metal, and 50 tons of uranium oxide. On December 2, with an audience of a Du Pont official and a number of project scientists, Fermi began methodically to remove the control rods. At each step he measured the neutron intensity and calculated its extent at the next stop. Hours later, with tension at a peak, he pulled the last rod out its prescribed distance, and the Geiger counters clicked faster and faster, becoming a steady buzz. Then Fermi gave the order to insert the control rods, and the chain reaction was stopped. [52:112]

Nuclear power had been produced, tamed, and halted for the first time. Some writers have hailed the significance of the event by calling it the opening of the atomic age, a sequel to the stone and iron ages. The amount of power was minute—only half a watt—but it would lead to the thousand-megawatt (thousand million, or 10^9, watt) reactors that utilities have today to generate electricity. It would first, however, lead to plutonium-producing reactors during World War II. CP-1 could run only for brief intervals, because the energy produced in fission shows itself as heat, which can quickly melt the components (as suggested on film in *The China Syndrome*, and as proven in reality in the Chernobyl disaster of 1986). Any reactor larger than CP-1 would require a heat removal system. (Most of today's reactors are, in effect, merely large water boilers that produce steam to spin turbines, with the water acting as both coolant and working fluid, and often also neutron moderator— unless graphite serves this last function.)

The Du Pont executive who witnessed Fermi's success was a member of a War Department reviewing committee whose job was to recommend whether a full-scale endeavor to build atomic bombs should be made. Two weeks earlier, the committee's draft report was negative. Now, the panel voted in favor of the effort. Two weeks after CP-1's debut, OSRD director Bush advised the president that full-size production plants should be constructed as quickly as possible. A gaseous diffusion factory for U-235 would cost $150 million and a reactor complex for plutonium $100 million, the numbers roughly a thousand times larger than at the pilot-plant stage. Roosevelt's approval on December 28, 1942, marked another important step in the march toward nuclear weapons: the United States was committed to trying to make nuclear bombs. [52:110–15]

Oak Ridge, Tennessee

The area around Oak Ridge had benefited from the electricity supplied by the Tennessee Valley Authority (TVA), but not enough to lift the Depression aura from the poor farms. However, once the army bought the large property and Groves's surveyors and laborers placed among the rolling hills railroad tracks, highways, streets, and a midsize city, as well as the uranium factories within a fenced military reservation, it was clear that the war had brought unexpected prosperity to the region. TVA's bountiful electricity was one of the reasons this location was chosen, and the large electromagnetic separation plant (code-named Y-12) and the gigantic gaseous diffusion plant (code-named K-25) would consume enormous quantities of it. Indeed, K-25 required more electricity than most American cities, for its thousands of motors and pumps had to move massive volumes of uranium hexafluoride gas through more than a thousand stages. Early in the planning it was decided that it had better have its own steam-electric plant, rather than place such a drain on TVA. [52:116–20, 130]

Specifications for the gaseous diffusion stages were exacting. Temperature had to be maintained precisely, for if lowered the gas could solidify, and if raised it would be more corrosive. Pressure, too, had to be constant, to avoid a pressure wave that could disrupt the process. The stages had to be made leak-tight, to prevent the loss of precious gas, to avoid human contact with the poisonous "hex," and to prevent the introduction of moisture from the air, which would encourage corrosion. Few industries relied upon large vacuum systems in their operation; scientists in the lab, perhaps more familiar with vacuums, were accustomed to the simple expedient of using grease and wax to plug leaks. Now, large equipment had to be made on assembly lines, to new standards of vacuum tightness. Research and development to improve production capability was still something of a novelty, but it was introduced in the Manhattan Project as well as in other critical areas, such as penicillin and rubber production. [52:124–25]

The outer structure that housed the gaseous diffusion equipment was four stories high, with a floor plan shaped like the letter *U*. Each leg of the *U* was half a mile long and 400 feet wide; it was the largest chemical engineering plant ever to be constructed as a single unit. It seems that the effort to separate or create quantities of special atoms and molecules required the biggest "this" or the most novel "that." In descriptions of the project, superlatives abound. Yet, while we may look with awe at this undertaking, the participants additionally dealt with dread. For although ground was broken for the K-25 plant in June 1943, its most critical component remained unavailable over a year later. The sieve, or barrier, through which the lighter isotopes would preferentially pass in each stage, resisted fabrication. Noncorrosive nickel was the material of choice, but the difficulties of preparing a nonbrittle barrier that could withstand the stresses and strains of installation and use, and that had uniform, microscopic pores of the desired size, were enormous. Even as the other components were installed in the plant, there was discussion of abandoning the gaseous diffusion process. But, true to the project's philosophy, they could not do that unless the method was proven impossible. Finally, using highly pure powdered nickel, there was the glimmer of success, and in early 1945 production of satisfactory barrier units rose. [52:123–28, 132–41]

The electromagnetic process also had its share of frustrations. Ninety-six vacuum-tight units were arranged as wedges fitting into an oval, called the "racetrack." Several racetracks were built, each 122 feet long, 77 feet wide, and 15 feet high. There was not enough copper available for the large magnet coils and bus bars, so tons of silver, an even better electrical conductor, were borrowed from the cache the U.S. Treasury had deposited at Fort Knox to back up the nation's currency. The operating cycle required each tank to be opened for removal of the product. This, of course, destroyed the vacuum, and the tank had to be pumped down for the next run. As mentioned above, the beams of the two isotopes were fuzzier than expected, requiring systematic scraping of the unit's interior to recover the desired material. A bomb demanded U-235 of at least 90 percent purity, but the electromagnetic plant yielded an enrichment of only about 11 percent. Groves used the gaseous diffusion product, which reached an enrichment of only a few percent, as feed for the racetracks; their output he then introduced into additional racetracks. In this patchwork way the necessary enrichment was achieved for the Hiroshima bomb. [52:149–67]

(U-235 for the first crude bombs was produced almost entirely by the electromagnetic method. Shortly after the war's end, the gaseous diffusion process reached its expected high operating efficiency, and this became the sole technique use by the United States for U-235 production. This may explain the First-World nations' surprise in 1991, when, in the wake of the Gulf War, it was learned that Iraq was endeavoring to make nuclear weapons by the outmoded electromagnetic technique.)

Hanford, Washington

With isotope separation so uncertain of success, the alternate route to a fission weapon was pursued vigorously. With data gathered from the operation of Fermi's original pile, another research reactor built at Argonne, and a pilot plant built at Oak Ridge, huge plutonium-producing reactors were planned. These "element growing" factories would operate at energies of hundreds of millions of watts, not the half watt of CP-1, and they would require a river of water to cool the uranium fuel. Elaborate instrumentation would be needed to control the reaction from a safe, shielded distance, and of course the plutonium would have to accumulate fast enough and in quantity sufficient for a bomb. Then it would have to be chemically extracted from larger masses of surrounding uranium.

For efficiency and to avoid problems with the intense radioactivity that would be released, the reactors had to be designed so the fuel could be unloaded easily after it had been irradiated long enough. Certainly, tearing apart a structure of graphite bricks each time would be impractical as well as dangerous. Engineers therefore designed piles of graphite bricks, with tubes to carry cooling water and with horizontal holes for the fuel; uranium would be pushed in at the front face and later unloaded at the rear.

With space and electricity at Oak Ridge inadequate for the reactors, Groves's staff searched for a relatively unpopulated region in western America, near a substantial river. They fixed upon the desert valley at Hanford, Washington, where the Columbia River sweeps a big bend and both Grand Coulee and Bonneville dams provided abundant hydroelectricity. Construction began in April 1943 of three water-cooled reactors, which were spaced for safety (including minimizing the danger of sabotage) at intervals of six miles. Two pairs of chemical separation plants were located at least ten miles from the closest reactor. Within months Hanford's boomtown resembled that at Oak Ridge, with 25,000 construction workers enduring primitive living accommodations and harsh climatic conditions. [52:188–90, 212–16; 92]

The Hanford factories also drew comparisons to the superlatives of

Oak Ridge. Colombia River water for the piles passed through a treatment plant large enough for the needs of a city of one million people. Each reactor building, a windowless box 120 feet tall, consumed nearly 400 tons of structural steel and tens of thousands of concrete bricks, concrete blocks, and cubic yards of poured concrete. Some 50,000 linear feet of welded joints had to be free of imperfections, for they would be inaccessible once the reactors were fired up. And in contrast to these massive images, the reactor base had to be set within a tolerance of 0.003 inches. At Chicago, Glenn Seaborg had determined the chemistry of plutonium from microscopic samples. This laboratory process was now scaled up a billion times in the design of the mammoth concrete "canyons" that housed the chemical separation apparatus. Intense levels of radioactivity would keep humans away from the units where irradiated fuel was chopped up, the aluminum jackets dissolved in acid, and the plutonium extracted from uranium and fission fragments. The process had to operate from afar; indeed, if a piece of equipment failed, it had to be disconnected from the system, unbolted, removed, and replaced with a new piece, all by technicians using remote controls. [52:216–22]

Just as the gaseous diffusion and electromagnetic separation plants of Oak Ridge were built before equipment and processes were perfected, similar problems of haste arose at Hanford. Water would flow over the fuel in each reactor and then back into the river, raising the temperature somewhat, but not enough to cause concern for fish or plant life. There was apprehension, however, about the potential harm from the radioactive fission fragments or uranium that might be picked up by the water. The simple solution was to encase the knockwurst-sized pieces of uranium fuel (called slugs) in some impermeable, heat-conducting material. Since the United States had vast experience in packaging food in cans, a similar process looked easy. Success nonetheless eluded the uranium canners for a year; as the reactors neared completion, there was no usable fuel. The problem was that the uranium cylinder had to fit tightly into its aluminum jacket, allowing proper heat transfer. Tiny pockets of entrapped air might swell and rupture the coating. The slug crisis was finally overcome by squirting molten aluminum into the can as the uranium was inserted, making a perfect bond; the lid was then welded on. [52:222–26]

Hanford's reactors were expected to be running in early 1945. The first stages installed in the gaseous diffusion plant at Oak Ridge were ready in the summer of 1944, while part of the electromagnetic apparatus was operating by that date. Yet, even if these massive production efforts were entirely successful, creation of a weapon was not assured.

Research and development on such questions as the size and shape of the supercritical mass (capable not merely of a chain reaction but of an explosive chain reaction), the means of detonating the bomb, the means of delivery, and explosion effects required yet another sustained endeavor.

Los Alamos, New Mexico

In September 1942, when Groves had assumed command of the Manhattan Project, he endorsed the idea that was floating around of building an isolated laboratory for the bomb makers. Security could be maintained around the fenced perimeter, while the scientists inside would be free to discuss all matters with each other. Few scientists supported Groves's extreme penchant for the compartmentalization of information. Their careers were built upon communication, for they relied upon gaining information from journals and colleagues. True, they competed with each other, but they knew that they gained more by practicing openness than by keeping quiet about their work. And when a project was completed, fame went to the person who attained priority by publishing first. Thus, not only in the Manhattan Project but throughout their careers, they saw positive benefits from knowing what the person in the next office was doing. However, engineers follow a different career pattern: secrecy. They succeed when they can patent a discovery before anyone else comes upon it.

To Groves, a remote laboratory would satisfy the working styles of the scientists and himself, and a location near a military proving grounds would be convenient for any large explosive tests that were needed. Additionally, he was disturbed by gossip circulating about the possibility of a hydrogen bomb and thought that isolating the scientists would help to keep a lid on it. The fusion of hydrogen can release enormous amounts of energy. The problem, however, was achieving the extraordinarily high temperature required for hydrogen to burn (which is why bombs based on this principle are called thermonuclear weapons). Fusion was an everyday occurrence in stars, but the only way seen to ignite a fusion burn on earth was by the temperature of a fission explosion. First things first, so fusion was placed on a back burner for the war's duration. [52:235]

Part of the Metallurgical Laboratory's charge was to explore the design of nuclear weapons. Toward this end, laboratory director Arthur Compton appointed J. Robert Oppenheimer in 1942 to head a fast-neutron study group. The members, like Oppenheimer, were theoreticians, most with long experience in nuclear physics, particularly the European emigrés Hans Bethe and Edward Teller. Oppenheimer himself had studied in Germany as part of a wave of young Americans in the 1920s, who sat at the feet of the masters creating quantum mechanics and then returned to bring physics in the United States permanently up to the "world class" level. Oppy, as he was widely called, split his time between the University of California at Berkeley and the California Institute of Technology in Pasadena and was a major force in establishing a strong tradition of theory in this country. Not by coincidence, for they shared the Berkeley campus and many physics problems, Ernest Lawrence was his counterpart in building a modern experimental tradition in the United States. [77]

Since a small bomb laboratory was at first envisioned (though by war's end it had grown to over 5,000 people), rail and road transportation and electricity and water supplies were of less importance than isolation. With encouragement from Oppenheimer, who often summered in the region, the site of a boys' school in Los Alamos, New Mexico, about 20 miles northwest of Santa Fe, was chosen by Groves. Though the more-than-a-mile-high mesa, with magnificent views of nearby mountain ranges, eventually proved to be too small, initially it seemed ideal for the work. Groves purchased the land by the end of 1942 and persuaded the business office of the University of California to act as the prime contractor under the Corps of Engineers in running the laboratory and its associated town. The army would provide security, while the university would act as employer of all personnel and buy whatever supplies were needed. [52:230]

Existing records do not explain why a California institution was selected to operate a facility in New Mexico. While it sounds unusual, it is not unique; astronomical observatories, for example, are built by some universities in distant countries as well as other states. Groves may have been impressed with Ernest Lawrence's "can do" enthusiasm and the support his administration gave to federal projects in the cyclotron laboratory and elsewhere on the Berkeley campus. Or, it may have been because support for Oppenheimer's fast-neutron group initially was routed through the University of California. Groves was impressed with Oppenheimer, who returned the respect.

Though Oppenheimer had no experimental experience and had never directed a laboratory, Groves apparently saw qualities of leadership in him—and probably more sympathy for Groves's style than was shown by most other scientists. Groves reviewed Oppenheimer's broad support of left-wing political causes in the 1930s and dismissed its significance; he was confident that Oppenheimer was not a communist and was loyal to the United States. (Military intelligence was less sanguine and kept Oppenheimer under surveillance.) Oppenheimer was appointed director of the Los Alamos Laboratory and soon convinced a stellar array of scientists to drop out of civilization until the war's end. Their contact with family (beyond spouses and children who joined them) and friends was through a postal box number in Santa Fe. Distinguished foreign-born physicists, including Bethe, Fermi, and Teller, senior Americans, and young tyros at the start of their careers filled the several working divisions of the laboratory, while consultants with special expertise came to visit. [52:227–39; 77; 101]

The grouping of human talent at all Manhattan Project sites is no less impressive than the remarkable construction accomplishments. These scientists and engineers were drawn away from other wartime enterprises, such as radar and the proximity fuse, and their willingness to gamble on nuclear weapons says much about their fear that Hitler would succeed first, as well as the excitement of working in close harmony with many brilliant peers and, for those moving the Los Alamos, their respect for Oppenheimer's abilities.

Construction on the mesa began in March 1943. Even before laboratory buildings were complete, apparatus from numerous universities was delivered. Harvard, Wisconsin, and Illinois all sent particle accelerators. Scientists boarded in the area's guest ranches until housing on the site was finished. The streets were muddy when it rained, the water supply became overtaxed, and the residents were pressed to create a town council to battle their army overlords on many urban problems. This common experience, plus the intensity of their work, created a strong feeling of camaraderie, and Los Alamos "alumni" often looked back to the period 1943–45 as the high point of their lives. [17]

Within the laboratory area a number of goals were set. Foremost, the scientists had to confirm that a chain reaction was possible with fast neutrons. Related to this was the need to know how rapidly new neutrons were released in each fission. Speed was essential to maximize the number of generations of fissions, and thus the energy liberated, before the bomb pieces were blown too far apart to react further. They also had to determine the quantity of scarce U-235 and plutonium needed for a supercritical mass; using too little would be pointless, while too much would be wasteful. Finally, and moving inexorably toward applied science, it was necessary to design the geometry of the weapons and plan their delivery. [52:240–43]

Since a stray neutron from the ever present cosmic rays might trigger a supercritical mass, it would be necessary to keep the U-235 and plutonium in subcritical condition until ready for detonation. The easiest way to accomplish this was to construct the heart of each weapon in two pieces and bring them together only at the time of explosion. Ballistics was a well-developed science, and plans proceeded rapidly for a gun barrel in which a projectile of uranium would be fired at a target of uranium; only together would there be a supercritical assembly. [52:245–46]

The gun method was unsatisfactory for plutonium, however, for this element was prone to fission spontaneously, and when alphas from the fission fragments struck impurities, they would add more neutrons. As a critical mass was assembled, the pieces would begin to react, then vaporize without much force. This danger of predetonation might be overcome by making the plutonium exceedingly pure, or by firing one piece at another at the extremes of muzzle velocity, but these solutions seemed impractical. Instead, the process of implosion was suggested. This technique, never before used, involved placing a sphere of high explosive around a small ball of plutonium. When the chemical explosive was ignited uniformly, a pressure wave would compress the center from all sides, squeezing the plutonium under enormous pressure into a much smaller and denser lump—a supercritical mass. If it would work, it had the added benefit of requiring less material. [52:245–49]

With much depending upon the rates of production at Oak Ridge and Hanford, Groves moved back Compton's schedule for the bomb's availability by half a year. It would be ready by the summer of 1945. Allied troops had landed on the beaches of Normandy, and it looked as if Germany would be defeated by that time. Apparently without any discussion or individual contemplation, the weapon conceived in fear of Hitler was planned for use on Japan. Though no one feared that the Japanese would construct nuclear weapons in World War II, they were a tenacious enemy and, to most people, it would have been incredible to withhold use of the new American bomb.

Homestretch to the Bomb

At the end of 1944, the Y-12 electromagnetic separation plant at Oak Ridge overcame most of its operating difficulties and began to produce substantial amounts of fairly pure U-235. Concurrently, the nearby K-25 gaseous diffusion plant approached completion as the individual stages, built in Detroit by Chrysler Corporation, were installed. Because the process could be initiated before the final stages were connected, Union Carbide Company, the plant's operator, was able to pump uranium hexafluoride gas into the primary units before the end of January 1945. By April the first enriched product was extracted. As mentioned above, the product of one process was used as feed for the other, and the bomb-grade material (over 90 percent U-235) was carried on trains by armed couriers to Los Alamos. [52:167, 294–302, 374]

The reactors at Hanford, Washington, seemed to reel from crisis to crisis. The problem of canning uranium slugs was solved in August 1944, but a more profound difficulty soon arose. The first reactor completed was given a fuel loading sufficient for operation, and the controlled chain reaction was carefully observed. All was satisfactory for a few hours, but then the reactor's power declined to the point of shutdown. Nothing of this sort had been predicted or seen in the experimental and pilot-plant reactors. Could the embarrassed scientists have missed a fundamental factor in the fission process? [52:304–8]

Soon they noticed the power rising, followed again by a shutdown. The cycle had a regular periodicity. This suggested that one of the many kinds of fragmentary atoms left after fission had an enormous appetite for neutrons. When enough neutrons were withdrawn from the chain reaction, the reactor turned itself off. But the fission product was radioactive, destroying itself with a half-life of about nine hours, and as it decayed to an insignificant amount, the chain reaction picked up, until more of the hungry fragment was created. The culprit turned out to be an isotope of a rare gas, xenon-135. There was no way to stop this isotope from being formed, but its effect could be overcome. With more fuel in the reactor, the chain reaction might proceed well in one corner while it was snuffed out in another. Fortunately, the Du Pont chemical engineers had insisted on a larger number of fuel slots than the physicists said were needed. For them, standard procedure was to build some flexibility into their plants. When all 2,004 loading holes in the face of the reactor were filled, the pile overall maintained a steady output. [52:304–8]

The other two reactors came on-line at the end of 1944 and in early 1945. As a Christmas gift to themselves, the Hanford pile operators withdrew several tons of irradiated fuel from the first reactor and forwarded it for processing in the chemical separation plant. A small amount of plutonium was successfully extracted. It was considered too risky to send it to Los Alamos by air, in case of a crash, and the train connections from Washington State were poor, so a military convoy of vehicles transported it. Huge factories at both Oak Ridge and Hanford turned out mere kilograms of product. Added to that apparent paradox, the vast majority of workers had no idea what they were making, aside from knowing that it was vital to the war effort. With many jobs no more interesting than turning a knob to keep a needle pointed to a certain number, it is remarkable that these plants were as efficient as they were. [52:308–10]

As the production plants increased their output of fissionable materials, attention focused more and more on Los Alamos. The U-235 gun-type weapon was such a reasonably straightforward job of science and engineering that it was accorded the highest confidence: no test was planned. The plutonium bomb was far more worrisome. With the strong likelihood of predetonation if a gun design was used, all efforts were concentrated on the implosion technique. Harvard chemical explosives expert George Kistiakowsky headed a new Explosives Division to work on this problem, and in addition to the American talent, from abroad came Niels Bohr and James Chadwick (as visitors), British hydrodynamics expert Geoffrey Taylor, and theoreticians Rudolf Peierls, Otto Frisch, and Klaus Fuchs. [52:310–13]

The crucial problem in implosion was to generate a uniform pressure wave toward the center of the spherical bomb. If the wave was asymmetrical, the plutonium core would not be squeezed into a spherical ball, the best shape for many generations of the chain reaction, but might squirt into an elongated shape of little use. The core, therefore, was surrounded by many sections of high explosive, which were cast into special shapes. Just as glass lenses of different shapes and materials are used to focus light waves, solid explosives of different kinds and forms can focus explosive (and implosive) waves. The shock wave thus shaped by the explosive lenses would exert enormous pressure upon the plutonium core, shrinking it into a supercritical mass. Los Alamos, conceived originally as a research laboratory, had to develop the casting and machining process to produce the lenses, as well as the electronic detonating circuitry, thereby becoming also a manufacturing facility. [52:312–16]

(As the twentieth century draws to a close, much concern has been expressed about renegade nations or terrorist groups obtaining nuclear weapons. The chief obstacles in their path are the acquisition of bomb-grade fissionable materials and the difficulty of mastering the implosion method. While the gun method can be used, it requires more material.)

A special Army Air Corps squadron equipped with the latest model bomber, the B-29, was formed in the autumn of 1944. Flying practice missions out of Wendover Field, Utah, they dropped dummy bombs of the projected size, shape, and weight. While reasonable accuracy from their bombing altitude of 30,000 feet was necessary, even greater attention was paid to the reliability of detonators, fuses, and the plane's release apparatus, and to the aerodynamics of the bombs' casings. [52:314]

The test everyone felt was necessary for the plutonium bomb was scheduled for July 4, 1945. What better day for a gigantic "firecracker"! Given the code name of Trinity, the explosion was planned for a section of the Army Air Corp's Alamogordo Bombing Range, about 100 miles south of Albuquerque. Harvard physicist Kenneth Bainbridge, appointed by Oppenheimer to direct the test, was responsible for erecting a 100-foot steel tower at ground zero and for the miles of wire connected to numerous instruments that would record the characteristics of the explosion. In preparation, he blew up 100 tons of high explosive, liberally doused with radioactive materials, to test the scientists' ability to monitor a radioactive cloud and its fallout. [52:318–19, 376–77]

Delays caused the test to be rescheduled for July 16. Even then it was touch and go, for cloud cover, rainstorms, and lightning threatened cancellation. The plutonium implosion device sat atop the tower, with detonators and firing circuit installed. At safe distances a small crowd of laboratory scientists and distinguished visitors watched from bunkers and from behind earthworks. Suddenly the darkness of the early morning hour was cut by a brilliant, yellow, warm glare. Once their momentary blindness dissipated, the astonished observers witnessed a swirling fireball of purple, yellow, orange, and red that expanded as it rose. A few seconds later the blast wave rolled over them, with the distant mountains echoing and reechoing the thunder. Even before the data were collected and analyzed it was obvious that implosion not only worked, but that it produced a larger explosion—the equivalent of about 20,000 tons of TNT (20 kilotons)—than most of the scientists expected. One week later, the plutonium core for the first true weapon (not a static tower device) was machined into shape, and a day after that the U-235 parts for the first gun-type bomb were ready, [44; 52:376–80]

Hiroshima and Nagasaki

Politics and Morals before Hiroshima

IN THE EARLY years of the project everyone was so busy creating the bureaucracy, laboratories, and factories and getting them up to speed that there was little time to reflect upon the larger picture. By mid-1944, however, with an Allied victory seemingly assured in another year or so, growing apprehension about the use of nuclear weapons in the current war and their role in the peace thereafter forced a number of people to reflect upon the "monster" they were creating. They were conscious of occasions when science was criticized, such as in the fictional Dr. Frankenstein's handiwork and poison gas warfare in Word War I, and were familiar with the movement, prominent primarily in Britain in the 1930s, that urged scientists to practice their profession with social responsibility in mind. They hoped that if nuclear weapons were to be used, it would be in a way that would minimize the loss of life.

Niels Bohr called upon Roosevelt in August 1944 to urge the internationalization of the atom. International control, he said, was necessary to avoid a postwar arms race. To create the proper cooperative atmosphere and show America's good faith, he urged a public announcement about the bomb project. While the object of these thoughts often was not spelled out, and the euphemism "international" used, postwar peace clearly depended upon the relationship between the United States and the Soviet Union. Bush and Conant agreed with Bohr, hoping that the president would make no agreements with Manhattan Project junior partner Great Britain that would raise problems for an accord with the Soviets. They too feared an arms race. But Roosevelt, in a meeting with British Prime Minister Winston Churchill in September 1944, indicated that he preferred to keep the project secret for the duration, and not to release freely to the postwar world technical information of commercial value, in order to give Britain's poor economy a head start. [52:325–29; 98]

Bush and Conant argued against this position, pointing out that international security was of primary importance, for any reasonably wealthy and technically advanced nation would be able to construct nuclear weapons. No monopoly on the raw materials was possible—even more so for hydrogen than for uranium—and scientists abroad could reproduce the American effort. It would be easier, in fact, since they would know from the start that it could be done. Share (even better, exchange) scientific information, not bomb-production technology, they urged, for secrecy would not enhance security. With openness, and with an international organization in charge of nuclear activities, perhaps other nations would not feel constrained to seek their own bombs. Especially after the Yalta conference in early 1945, when it appeared that it might be possible to cooperate with Joseph Stalin's huge empire, the chance to avoid a weapons race seemed too important to lose. [52:329–31]

Leo Szilard, always in the forefront of political as well as scientific currents, had encouraged the Met Lab as long ago as September 1942 to discuss the political dimension of their labors. Ineffective then, he was again unsuccessful in March 1945, when he implored Einstein to send another letter to Roosevelt; the president died before reading it. His successor, Harry S Truman, who like most vice-presidents in this country was kept "out of the loop" regarding important matters, soon learned of the uranium and plutonium explosives from Groves and Secretary of War Stimson. With the war in Europe nearing an end, and with no special consideration that the weapon had been conceived in fear of Germany making one, the cities of Japan slid easily under the bombsight. [52:342–43]

Germany surrendered to the Allies on May 8, 1945. Truman looked forward to a meeting with Churchill and Stalin in July, in the Berlin suburb of Potsdam. The conference's purpose was to decide questions about the political and economic character of liberated Europe and to discuss the war in the Far East. Truman planned also to discuss nuclear weapons with Churchill and perhaps with Stalin. As part of his preparation for Potsdam, Truman appointed a high-level group of officials, called the Interim Committee, to advise him on all matters relating to the new weapon. The Interim Committee in turn named a group of four scientific advisors: Arthur Compton, Enrico Fermi, Ernest Lawrence, and Robert Oppenheimer. [52:344–46]

To protect its far eastern flank in the early days of its conflict with Germany, the Soviet Union had signed a nonaggression pact with Japan. However, the other three Big Four powers, the United States, Britain, and China, were belligerents with Japan and demanded unconditional surrender of their enemy. Such a demand, while useful for domestic political purposes, can make problems in terminating conflict, for most surrenders do involve some conditions. For the last few years the United States had asked the Soviet Union to declare war on Japan, reasoning that this would tie down millions of Japanese troops in Manchuria, leaving fewer to defend the home islands against an Allied invasion. In return for some Japanese-occupied territory on the Asian mainland, including land captured in the Russo-Japanese War, Stalin agreed to attack the nation that had humiliated Russia in 1905. At the Yalta conference he set the date as two or three months after the defeat of Germany, to allow time to transport troops eastward on the trans-Siberian railway. [52:348–53]

During the spring of 1945, some of the Yalta pledges seemed to unravel; the United States and the Soviet Union had different interpretations of their agreements. The Soviets, for example, established satellite governments in Eastern European countries, where the United States and Britain had expected truly free elections to be allowed. With irritation turning to hostility, the United States began to rethink its Pacific scenario. It might be simpler to conclude the war against Japan without the Soviet Union's uncertain assistance. Indeed, a joint occupation of Japan, as was the situation in Germany, was regarded with apprehension. [52:348–53]

Besides the varying viewpoints on the need for Soviet help and the utility of the unconditional surrender formula, the American military leadership was not of one mind about the strategy to be followed against Japan. Presidential Chief of Staff Admiral William Leahy and Chief of Naval Operations Admiral Ernest King favored naval blockade and saturation bombing of the home islands to defeat the Japanese nation, while Army Chief of Staff General George Marshall, supreme Allied commander in the Pacific General Douglas MacArthur, and Admiral Chester Nimitz believed that an invasion, despite heavy initial losses, would entail fewer casualties overall. A large uncertainty was whether Japanese civilians would fight to the death if unconditional surrender meant dethroning their emperor. In May the Joint Chiefs of Staff fixed November 1, 1945, as the date for an invasion of the island of Kyushu. The yet untested atomic bomb apparently played no part in these plans. [52:348–53]

The Interim Committee was given a broad mandate: to discuss postwar legislation and organization, research, civilian applications of nuclear energy, and international control, in addition to the wartime use of the atomic bomb. But it was the last issue that absorbed them as they met several times in May. Committee member James Byrnes, soon to be Truman's secretary of state, was influential in defining the U.S. position that technical information not be shared with the Soviet Union. The scientific advisors examined whether a harmless demonstration of the bomb could be staged that would nonetheless lead the Japanese to surrender. Inviting some Japanese leaders to view an explosion in a deserted place was considered unlikely to convince their government in Tokyo. A trial detonation over a prenamed, scarcely populated part of their homeland was too risky: the test might fizzle and the fissionable material be recovered; Japanese fighter planes might attempt to shoot down the bomber; or prisoners of war might be brought to the target. [52:353–60; 98]

Finding no spectacular way to demonstrate the bomb safely, the advisors and the full committee concluded that a surprise attack on a military or industrial target surrounded by workers' homes would have the greatest psychological impact. Choosing an urban objective was not unusual at that time, as American and British bombers had flown against numerous enemy cities, including incendiary raids on Hamburg, Dresden, and Tokyo, while German V-1 and V-2 rockets had been launched against England. Indeed, the Interim Committee, meeting before the unexpectedly great success of the Trinity test, could not know if a nuclear weapon would destroy more of a city and kill more civilians than a conventional air raid. [52:353–60]

The Interim Committee recommended further that the Soviet Union not be informed of the new weapon before it was dropped. Things would be awkward if they insisted on a voice in its use. Looking to the future, the committee advocated complete openness by all countries with respect to their nuclear work, backed by an international agency with the power to inspect facilities. Herein lay the seeds of a controversy that spanned the next 45 years: the United States, unwilling to rely on trust to maintain adherence to arms control treaties, insisted on verification of the behavior restricted by the treaties; the Soviet Union, seeing every attempt at foreign inspection as a cover for spying, adamantly rejected such scrutiny. [52:360–61; 98]

At about the same time that these political questions were debated, the military and scientists selected the targets for the bomb. Kyoto, Niigata, and Hiroshima were chosen, in part because they were industrial centers (Hiroshima also contained an important port and an army headquarters) and in part because U.S. long-range bombers had not yet attacked them. Nuclear weapon effects could therefore be distinguished clearly, and their psychological impact would likely be greater. Because of its historical and cultural significance, Secretary of War Stimson deleted Kyoto, and Kokura was placed on the list; Nagasaki was later added as an alternate. [52:365]

Scientists in Chicago knew that nuclear policy was being formulated in the spring of 1945, and they were disturbed that people at the workbench level were not being consulted. Research at the Met Lab had slackened, once the fruits of its labor were transformed into concrete and steel at Hanford and Oak Ridge. Groves provided some funds for a heavy-water reactor and other experiments but really had no authority to press forward on a postwar program. For those who felt that nuclear research should be pursued vigorously, to maintain the country's leadership—because safety lay in continued research—it was a depressing time. Perhaps because they had little scientific work to do, they put their minds to the political ramifications of the birth of the bomb, and they wanted to be heard. Met Lab director Compton sympathized with this need. As an advisor to the Interim Committee, he felt that he could serve as a conduit and therefore encouraged his staff to prepare position papers on a range of issues. [52:365–66]

The best-known document, from the committee dealing with social and political problems, emerged in June. Taking its name from the chairman, refugee physicist and Nobel laureate James Franck, the Franck Report argued that neither secrecy nor an attempted monopoly of the raw materials could prevent a nuclear arms race. Indeed, their reasoning went, in such competition the United States would be at a disadvantage, since its population and industry were concentrated in fewer urban areas, which could be attacked with fewer bombs. International control was the key to the solution, but to attract and hold the goodwill of other nations the United States must not use the new weapon on Japan without warning. [52:366]

At a meeting in late June 1945, the Interim Committee debated the Franck Report and others from Chicago. Since the scientific advisors still could think of no benign but effective demonstration of the bomb, and the drawbacks to an announced use already mentioned still pertained, the committee continued to recommend bombing a military-industrial target without warning. When the committee reexamined its position on informing the major US allies about the bomb, however, it changed its mind. Britain, as a Manhattan Project partner, had to be told, while China and France were of little concern; clearly, the Soviet Union was the focus of discussion. Now the committee advised Truman to tell Stalin of the bomb at the forthcoming Potsdam conference. Should the Soviet leader ask for more information, the president could say it was unavailable. The merit of this charade was that the United States could say it had taken its ally into its confidence. [52:367–69; 76; 98]

At Potsdam in mid-July, the Soviet Union largely rebuffed British and American concern over the puppet governments it was installing in several Eastern European countries. It was obstinate also in demanding further territorial concessions from China before it would declare war on Japan, though by this time Truman would have been glad to go it alone. Moreover, the United States, having broken the Japanese diplomatic code early in the war, knew that Tokyo was endeavoring to have Moscow act as an honest broker in reaching a peace settlement with the Allies. But the Soviets were noncommittal to the Japanese and silent to the Americans. Truman and Churchill, further, had to wrangle with the Allied insistence on unconditional surrender. Retracting this demand might have devastating political repercussions at home, but keeping it might result in a fanatical defense by the Japanese, who feared deposition of their emperor and dissolution of their nation. [52:380–86; 76; 98]

While at the conference, Truman learned of the successful Trinity test of the plutonium implosion device. Stimson urged that Japan be warned to surrender, but Secretary of State–designate Byrnes counseled that, if they refused, angry American voters might turn against Truman's party at the next election. Once again the tail of domestic politics wagged the dog of foreign policy. [52:386; 76; 98]

Truman followed the Interim Committee's suggestion that he inform the Soviets about the bomb. It must have been the most low-keyed diplomatic presentation of the century. After a session concluded, Truman casually walked around the table and unceremoniously mentioned to Stalin that the United States had developed a new weapon of unusual power. Stalin seems to have exhibited no interest and merely said that he was pleased to hear this news and hoped that it would be used well against Japan. For many years it was unknown whether Stalin comprehended Truman's message. In 1969, however, in Marshal Georgi Zhukov's memoirs, it was revealed that Stalin understood precisely what weapon Truman meant, for Soviet spies had penetrated the Manhattan Project. Stalin quickly increased the tempo of his own project and placed secret police chief Lavrenti Beria in charge. [52:394; 58; 76; 98]

The Potsdam Proclamation, signed by the United States, Britain, and China (the Soviet Union not being a belligerent) and issued on July 26, 1945, cautioned Japan of "prompt and utter destruction," but it said nothing of any new weapon. Japan was assured a peaceful government selected by the people, but the document was silent about the emperor's status. America knew that the peace faction in Tokyo sought acceptable terms for surrender, but this proclamation contained little support for them. The hard-liners could maintain that the Allies intended to destroy the throne, which was the core of national identity. Tokyo thus responded to the Potsdam Proclamation with a word that was translated into English as "rejection," though in the original Japanese it was far more ambiguous. [52:395–96; 76]

Hiroshima and Nagasaki

None of these political discussions or events slowed construction of the bombs. Indeed, the pace quickened as Los Alamos scientists endeavored to meet the schedule Groves had set some time before. There was pressure, too, from the White House, as Truman wanted a "big stick" in his hand when the neophyte diplomat sat down in Potsdam with "old pros" Stalin and Churchill for sessions of hard bargaining. As historian Martin Sherwin well understood, "the decision to use the bomb to end the war could no longer be distinguished from the desire to use it to stabilize the peace." [98:221]

The first weapon was expected to be ready by early August 1945. In anticipation, the special bomber squadron flew to the large American air base on Tinian Island, in the Marianas. Toward the end of July, the cruiser *Indianapolis* delivered the U-235 gun and part of the fissionable material, while the rest of the U-235 arrived by air; within a short time the bomb was assembled. Because visual and photographic impressions of the explosions were desired, radar-guided drops were unacceptable. When cloudless skies finally came on August 6, weather planes flew some 1,500 miles to three of the cities, radioing back that the primary target, Hiroshima, was clear. A formation of three B-29s—the weapon plane, the photography plane, and an instrumentation aircraft—turned toward the island of Honshu.

The planes approached the city at an altitude of 30,000 feet in order to keep as much distance as possible between them and the blast effects. After the *Enola Gay* released the uranium bomb, it banked sharply into a turn also designed to minimize any danger from the shock wave. At 0815 the bomb exploded with a piercing flash of light over Hiroshima, followed by a large fireball and then the boiling, malignant, expanding cloud of flames and smoke. As the cloud rose to 40,000 feet, it trailed behind a thinner column, the entire image now familiarly called a mushroom-shaped cloud. Observers in the aircraft saw numerous fires throughout the city, though a pall of smoke obscured any clear picture. The explosive force was estimated to be about 20 kilotons (recalculated to 12.5 kilotons many years later). [52:401–2; 88:708–11]

Truman, on the high seas returning to the United States, broke the news to a startled world, warning the Japanese of further destruction. In Tokyo, the peace faction in the cabinet urged the emperor to accept the Potsdam Proclamation, but, with little hard information about the attack and its damage coming from a devastated Hiroshima, the cabinet's majority counseled inaction. The Soviet Union now told the Japanese ambassador in Moscow that their nations would be at war on August 9. Some saw this as a cynical move to be able to claim the spoils of war before the bomb ended it and preempted such a right, but the Soviets argued that they were merely living up to their promise to join in the Pacific conflict within three months after Germany surrendered. Faced with this new misfortune, and with six million leaflets dropped by American planes urging the Japanese to surrender, the cabinet remained deadlocked. [52:403–4; 88:737]

On August 9, 1945, aircraft sought to repeat the military success of the attack on Hiroshima three days earlier. But this time the photography plane failed to rendezvous with the others. They circled nearly an hour, consuming precious fuel, while the clouds grew thicker over Japan. By the time they reached the primary target, Kokura, the city was obscured by haze and smoke. After three passes, they flew on to Nagasaki. With fuel only for one bombing run, *Bock's Car* approached by radar, plunged through an opportune hole in the clouds, lined up on its target, and dropped the plutonium implosion weapon. The plane limped to a friendly field closer than Tinian, landing with only fumes remaining in its gas tanks. It left behind 20 kilotons' worth of demolished city. [88:740]

The damage to these two cities was awesome. In Hiroshima the explosion occurred at a time when the breakfast stoves still held hot coals. The pressure wave leveled many poorly constructed houses, yet the blast itself ignited few fires. Thousands of overturned stoves, however, kindled numerous fires that easily spread to flammable residential debris, creating a huge inferno called a firestorm. A rising column of hot gases sucked in air from the circumference at maximum velocities of 30 to 40 miles per hour. (Firestorms caused by incendiary bombs had also destroyed much of Tokyo, Hamburg, and Dresden, in some cases consuming so much oxygen that many people died of suffocation.) In an instant, the infrastructure of the city was destroyed. Police, fire, and military organizations were useless; many of their personnel were injured, while the debris-filled streets made it difficult for the able-bodied to move about. Eighty percent of the firemen and 70 percent of their equipment were placed out of action; the water mains were crushed anyway, so the fires burned themselves out. Medical personnel were casualties also, and hospitals were damaged. In the urban area, 62,000 out of 90,000 buildings were destroyed; reinforced concrete structures fared the best, but they were largely gutted. In an instant, a functioning city was no more. The disorganization was almost total; significant help could come only from outside. Except for the lack of a firestorm, Nagasaki experienced similar devastation. [74; 106]

Statistics cannot portray the horror of these nuclear disasters, but they can convey an impression of the extent of the damage and injuries. There are no authoritative data, for no one knew the exact populations of the cities; US figures tend to be lower and Japanese higher. The numbers presented here are therefore approximations, taken in large part from the US Strategic Bombing Survey report. For comparison, statistics are given for the results of the 1,667 tons of incendiary bombs dropped on Tokyo the night of March 9–10, 1945.

	HIROSHIMA	NAGASAKI	TOKYO
Population at the time	250,000	195,000	???
Dead	70–80,000	35–40,000	<100,000
Wounded	70–80,000	35–40,000	<80,000
Area destroyed in square miles	4.4	1.8	15.8
Population density per square mile	35,000	65,000	130,000
Deaths per square mile	17,000	21,000	6,000

Despite Tokyo's greater population density and area destroyed, the "intensity" of death was less than at Hiroshima and Nagasaki. Nuclear weapons pack a larger "punch." Nagasaki suffered fewer casualties than Hiroshima, and a smaller region was demolished, because the blast was confined to the industrial valley; yet, due to its relatively high population density, it had the highest mortality density. [74; 106]

These two nuclear attacks crushed the regional army headquarters and port in Hiroshima and the Mitsubishi shipyards, electrical equipment works, arms plant, and steel works in Nagasaki, as well as the cities in general. They did still more. They were psychological weapons designed to crush the will to resist and, further, to give the Japanese military a means of honorable surrender. There would not be the shame of losing to superior Allied forces; they could not be expected to resist nuclear weapons. How large this concept loomed in the minds of the Japanese leaders is impossible to say, but it likely played some role.

Nuclear weapons, such as those that leveled Hiroshima and Nagasaki, differ from conventional weapons, such as those dropped on Tokyo, in the intensity of destruction that they cause and in still other ways. They are not like ordinary chemical explosives "scaled up" to larger size. They are qualitatively as well as quantitatively different. Aside from the radiation they emit, which sets them apart from usual explosives, such weapons have the ability to achieve a "saturation attack." This means that there is nothing left to justify returning to the target (unless it is a buried, reinforced bunker or missile silo). Churchill aptly described a further attack as "making the rubble bounce."

According to Japanese protocol, government actions were taken in the emperor's name, but decisions were actually made by the cabinet and presented to the ruler for his customary approval. Emperor Hirohito violated this tradition by insisting to his cabinet that the nation had suffered enough and the fighting should end. Word was sent through neutral countries to the United States on August 10 that Japan would accept the Potsdam Proclamation, as long as the emperor was allowed to continue to rule. Washington replied that the emperor would reign, but under the control of the occupation forces. Even though hardly unconditional surrender, Japanese military leaders rejected the terms. Once again, Hirohito declared that the conflict must end. On August 14, 1945, the Japanese cabinet finally accepted the Allied terms, and World War II was over. [52:404–5; 82; 105]

Reaction to the Bomb

Amid the euphoria of victory, there developed a wide range of reactions to the bombing of Hiroshima and Nagasaki. Ambivalence, uncertainty, and agony were reflected in Oppenheimer's now famous remark to an audience at the Massachusetts Institute of Technology (MIT) in November 1947: "In some sort of crude sense which no vulgarity, no humor, no overstatement can quite extinguish, the physicists have known sin; and this is a knowledge which they cannot lose." [81] Horror was expressed much closer to the bombings, in the summer of 1945. *Christian Century* ran an editorial called "America's Atomic Atrocity," while *Commonweal* published an article entitled "Horror and Shame." [4; 59] *Time* magazine quoted one man's condemnation, to the effect that it is a "barbaric, inhuman type of warfare." Another asserted that "it is simply mass murder, sheer terrorism." Yet another said, "The United States of America has this day become the new master of brutality, infamy, atrocity." And another: "Man is too frail to be entrusted with such power." [80] John Foster Dulles, within a decade to be Eisenhower's secretary of state and noted then for his aggressive willingness to go to the "brink" with the Soviets, was of a different mind in 1945. In a statement on behalf of the Federal Council of Churches, he said ruefully: "If we, a professedly Christian nation, feel morally free to use atomic energy in that way, men elsewhere will accept that verdict... the stage will be set for the sudden and final destruction of mankind." [80:36]

But such voices were in the minority. Just one week after the above-mentioned article, *Commonweal*'s editor denounced the lack of criticism of the bomb's use. [59] Indeed, most Americans felt proud of their military and their scientists and relieved that the conflict had ended. If the Japanese suffered the obliteration of two cities, widespread opinion felt that they deserved this for their aggression. The cry "Remember Pearl Harbor" faded slowly from memory (indeed, it echoed half a century later, on the fiftieth anniversary of the attack, in 1991, but then in part as protest of Japan's economic power). At the end of 1945, *Fortune* magazine published a survey showing that under 20 percent of the respondents held any moral reservations, while over 50 percent believed the United States had behaved properly. An *additional* 23 percent were sorry that the war ended before more fission bombs could be dropped on Japan! [41] Such sentiments might have been reinforced had the American public known at that time of the significant, but mismanaged, German bomb project and the much smaller Japanese nuclear effort, both of which failed in their goal. [107]

In the years since, historians, politicians, military figures, and scientists have examined and reexamined the bombing of Hiroshima and Nagasaki. Was it necessary to ensure Japan's surrender? Was it moral? How would the world be different if the bombs had not been dropped? Despite heated debate, definitive answers are not possible. These are, after all, questions involving predictions and value judgments; individuals will differ, as they will on any controversial issue. Still, examination of such questions does lead to greater insight into the political process.

Though no longer able to wage aggressive warfare by the summer of 1945, Japan had a few million men under arms on the home islands and additional forces on the Asian mainland. These troops had caches of ammunition and many kamikaze aircraft available for their suicide missions against American warships. After three and a half long years of war, the United States had no intention of stopping short of occupying Japan and removing its government. Domestic political realities would allow nothing less. An invasion was expected to be costly, as suggested by the determined resistance to the American landing on Okinawa. Given these circumstances, as well as the Japanese unwillingness to surrender unconditionally, and the belief that total casualties would be fewer if the war ended sooner, why not use nuclear weapons to shock them into capitulation? [38]

The reasons "why not" are based largely on the national belief that Americans are a moral people. We want to feel that we do the right thing. Yet, in this case, there were a few notes of discord even before the bombings and many more afterward. Secretary of War Stimson, who never wavered in his intent to use it, nonetheless called the bomb "the most terrible weapon ever known in human history." Scientists, such as those at Chicago, warned of undesirable consequences: revulsion at the US action and a postwar arms race. General Dwight Eisenhower, supreme Allied commander in Europe, felt that bombing a defeated Japan, even if it had not yet surrendered, was unnecessary and might shock world opinion. General MacArthur believed that the bomb was unnecessary. General Marshall expected the Soviet declaration of war against Japan and a proposed change in the unconditional surrender formula to be more significant than the bomb. Admiral Leahy disdainfully likened atomic bombs to germ warfare. "The use of this barbarous weapon at Hiroshima and Nagasaki was of no material assistance in our war against Japan," he said. "Wars cannot be won by destroying women and children." Less chivalrous, perhaps, and more practical, Admiral King, General Carl Spaatz, commander of the Strategic Air Forces, General H. H. Arnold, commanding general of the Army Air Corps, and other military leaders opined that naval blockade and conventional strategic bombing would have ended the war quite soon without an invasion and without nuclear weapons. While it is possible that these officers' views were tinged a bit by "sour grapes," for they had to share their truly impressive victory with the scientists, their rejection of the bomb's need is hard to dismiss entirely. [1; 2; 19:12; 34:312–13; 40:31; 79]

Some diplomats also expressed reservations. Joseph Grew, who had been ambassador in Tokyo for ten years and was undersecretary of state in 1945, knew from intercepted diplomatic messages that the Japanese were feverishly seeking an honorable means of surrender. He tried to have included in the Potsdam Proclamation that the emperor would keep his throne but lost to the arguments of James Byrnes that this would be regarded as revising the concept of unconditional surrender, something unacceptable to the American public (who happily accepted the reality of conditions a month later) and a sign of weakness in Japanese eyes. [40:32]

Members of the US Strategic Bombing Survey that studied the issue after the war early concluded that "certainly prior to 31 Dec. 1945, and in all probability prior to 1 Nov. 1945, Japan would have surrendered even if the atomic bombs had not been dropped, even if Russia had not entered the war, and even if no invasion had been planned or contemplated." [40:87] With the 20/20 vision of hindsight, and, even more significantly, through the eyes of people at the time, the military reasons for using the bomb do not seem overwhelmingly persuasive. Whey then was it used?

The answer, for some, was international power politics, also called nuclear diplomacy. Perhaps the first to level this charge was physicist Patrick Blackett, a British Nobel laureate who, during the war, helped found the science of operations research, applying it to maximize convoy defenses against submarines. In 1946 Blackett argued that the United States rushed to use nuclear weapons in order to thwart the Soviets from any significant role in defeating Japan and to make them more receptive to Allied wishes in Europe. The bombings of Hiroshima and Nagasaki were, he said, not the last operation of World War II but the first act of the Cold War. [25; 79]

Historian Gar Alperovitz breathed new life into these views about two decades later. He supported Blackett's thesis but changed the primary focus of U.S. interests. The nations of Eastern and Central Europe were of far greater concern to the United States than any parts of the Asian mainland that the Soviets might try to dominate. Alperovitz was persuaded that power politics were at the root of the situation, for the United States failed to respond to Japanese peace feelers, ignored domestic advice against using the bomb, and declined to wait long enought to see the effect on Japan of the Soviet Union's declaration of war. These charges were, of course, controversial, and most historians were not persuaded by them. It was hard to believe in a conspiracy by the president and his secretary of state to destroy two Japanese cities primarily as a gesture against the Soviet Union. [1; 19:65; 79]

Other players joined in the controversy. Allied soldiers scheduled to storm the beaches of Japan, and prisoners of war, were convinced that the bombings saved their lives. Communists and others who looked upon America an an imperial power were convinced that the United States behaved cynically. Peace activists sought to establish a moral precedent against using such weapons in the future, while hard-liners fought for the opposite stance.

A centrist position that recognizes some validity in all the arguments may come closest to reality. Decisions are rarely based upon a single motive; life is too complex for that. In this case, the military justification of fewer total casualties meshed comfortably with the American penchant for "wrapping up" the war; we are not a patient people. Added to that were the domestic political gains of using a most impressive invention, versus the political embarrassment of withholding a weapon produced at great expense that might have saved some American lives. International politics were mixed in as well. The bomb kept the Soviet Union from joining in the occupation of Japan, and the American president for a while could hold a club over Stalin, hoping to restrict Communist expansion. If Truman made a list of the pros and cons, it seems almost inevitable that he would have decided to use the bomb.

There is a further wrinkle, however. The Manhattan Project developed tremendous momentum by 1945, and virtually all involved assumed and expected that nuclear weapons would be dropped on the enemy. Opposition came too late to make much of an impact. Seen in this light, the president was merely asked to affirm long-standing policy; he was not really required to make a decision. The only true decision that could have been made would have been *not* to use the bomb. With hindsight, this may seem to have been a desirable option, and even at the time Truman was counseled to reconsider its planned use. But the president, stretching his newfound geopolitical muscles, lacked the psychological or moral resources to consider that option seriously. Then again, one might regard it as remarkable that there was *any* debate over nuclear weapons; there was virtually none about deadly fire raids and blockades, both of which were also aimed primarily at civilians.

Were such political and moral issues appropriate concerns for Manhattan Project scientists, who, after all, lacked training in these subjects? They thought so, both as citizens and as individuals who had given the problems informed attention. But they were in a curious position. They were regarded with awe, for these men-who-made-the-bomb were seen to be in intimate, daily contact with the most profound secrets of nature. Yet they were forbidden by security laws from saying much publicly about those secrets. Then the Smyth Report, the official army description of the project, was published, containing such a detailed description of the processes and personnel that the hunger for factual information was sated. Scientists thereupon turned their attention to two things: the political ramifications of the atomic age, and their own employment. So many left the project sites for both old and new academic positions that the army was concerned about its ability to improve the weapon further and build a stockpile. Oppenheimer, for example, returned to Berkeley and not long thereafter accepted the directorship of the Princeton Institute for Advanced Study. Fermi, Szilard, and others by this time had been on the staff of the University of Chicago for so long that they accepted permanent positions there.

Political ramifications concerned not only the scientists. Newspaper columnists Joseph and Stewart Alsop approvingly quoted an army policy study that concluded: "The only sure defense of this country is now the political defense." They advocated international control by the United Nations. [3] Hanson Baldwin, military affairs writer for the *New York Times*, argued that offensive actions would now always conquer defensive ones; the United States must therefore maintain its edge in research, development, and mass production. [20] Scientists, if they did not dominate this discussion, were a very visible part of it. In a best-selling anthology called *One World or None: A Report to the Public on the Full Meaning of the Atomic Bomb*, Compton, Einstein, Oppenheimer, Szilard, and others spoke out against secrecy and for international control of nuclear energy. [75] None of these viewpoints was expressed casually; all were part of a great debate on the future direction of the nation and the world. The issues involved the structure of military forces, soon to be termed an arms race; the extent and wisdom of civil defense; the nature of domestic management of the atom, before long placed in the hands of a newly created Atomic Energy Commission; and whether the infant United Nations could be the agency to control the atom internationally.

The New World

Domestic Control of Nuclear Energy

IN HIS EARLY remarks about the bomb, Truman had casually mentioned its "secrets." Scientists were discomforted by this term, for they knew that information about nature that was discovered in one country could certainly be found in another. Even real secrets—such as the details of fissionable material production and the technology of the weapons—would fall to a determined assault, and thus could no more assure a monopoly on power than could an aborted plan Groves had devised to corner the world supply of uranium. In their minds the American public accepted these rational arguments, but in their guts they felt fear and thus sought to protect "our" secrets. Consequently, secrecy and national security became synonymous. How then could security be achieved? By controls. Domestically, this meant legislation; internationally, it meant a treaty. In both cases secrecy was never far from center stage.

Oak Ridge physicist J. H. Rush, a keen observer of the scene, noted that "when the late war ended in a thunderclap, it left two noteworthy developments in its wake. Science had become politically interesting, and scientists had become interested in politics." [99:ii] These were distinct phenomena, but the lines were blurred because both were directed at atomic energy, and both reached a crescendo in the year between the summers of 1945 and 1946. It was a period when the public relived the horrors of the war through such events as the Pearl Harbor investigation, revelations of the atrocities at Auschwitz and Bataan, and the trials of war criminals. It was also a time to ponder the hopes and fears of the future, represented by the new United Nations organization, labor difficulties, the return of troops from abroad, and ever-increasing discord with the Soviet Union. The bomb tapped into both wartime and postwar concerns. During the conflict a wit remarked that if the Manhattan Project was successful, Congress would not even

63

be curious about the expenditure of $2.2 billion, but if it failed, Congress would investigate nothing else. In fact, Congress became absorbed with the triumphant venture and spent much of the next year debating the Atomic Energy Act of 1946, which created the Atomic Energy Commission (AEC). [99]

Attorney Byron Miller, who participated in the legislative battle, indicated just how novel the situation was for the participants:

> Several unique factors combined to deprive the legislator of his comfortable patterns for reaching policy decisions. He was not dealing with a recast of conventional controversy, a labor versus management or debt reduction versus public spending issue, on which his attitudes had long been fixed, his speeches ready at tongue, the public reception and opposition tactics already known. He could not judge by the people lined up on one or the other side, for traditional alignments were criss-crossed. Even commercial special interest groups were largely silent. In short, for most senators and many representatives, atomic energy legislation required an almost pure exercise of judgment. [78:799]

If Congress was intimidated by the responsibility to craft a law, the scientists were outraged. To their astonishment, on October 3, 1945, a bill written by the War Department for the Interim Committee was introduced into the House by the chairman of the Military Affairs Committee and into the Senate's corresponding committee by its ranking Democrat. Most scientists had deliberately been kept uninformed that such a bill was being drafted. Within a week, the House committee held merely a single day of public hearings and then moved into closed sessions to finish the job quickly. [52:428–55, 482–530; 78]

The secrecy and high-handedness of these actions were ill considered tactically, strategically, and democratically. Although the War Department had designed legislation that created a civilian agency, the May-Johnson bill (named after its sponsors) was widely denounced as a covert grasp for military control of nuclear energy. This seemed the intent because the bill created a part-time commission that could easily be dominated by the full-time administrator and deputy administrator. In a departure from tradition, appointment to the full-time positions of active duty officers was permitted—and expected. Only military policies were outlined for the new agency; support for basic research, declassification of information, and civilian applications went unmentioned. The agency, further, would license research and could even bar scientists in nongovernmental laboratories from such work. The widespread perception of military manipulation did nothing to slow the hemorrhage of scientists from Manhattan Project sites.

Action in the Senate on May-Johnson was more deliberate. Indeed, a jurisdictional problem had to be resolved before they could proceed: which committee would receive the bill? Legislators usually seek high-visibility assignments, for these can translate into votes at the next election. Thus, there was no lack of committees anxious to share headlines with the legislation.

The scientists already had formed the Atomic Scientists of Chicago, the Association of Oak Ridge Scientists, and other groups. Soon these coalesced into the Federation of Atomic Scientists, and before long this merged with the new Federation of American Scientists (FAS). This infrastructure, tentative and unsophisticated as it was, issued news releases and directed an educational campaign of impressive proportions. The *Bulletin of the Atomic Scientists,* begun as a newsletter for the Chicago group, soon became the premier national journal for science and public policy. Aimed as much at the layman and politician as at the scientist, the *Bulletin* initially concentrated on atomic energy topics but later expanded to questions of world health, food supply, population control, space, and other matters of policy with a technical component. Like the FAS, the *Bulletin*'s expertise still commands a respectful audience. [78; 99]

For that audience in 1945, the "scientists' movement" was happy to dissect the May-Johnson bill. It pointed out that research would be hindered by the bill's restrictive provisions; this would not be in the best interest of the nation. But the movement did not just analyze the proposed legislation. When bare-knuckle politicking was needed, the scientists used the simplistic but effective rallying cry of "civilian control of atomic energy!" [78; 99]

A few prominent scientists were active in this movement—Leo Szilard, Edward U. Condon, and Harold Urey among them—but the majority were idealistic young men who had yet to make their reputations. The furor they created for an opportunity to present their views gained only a single day more of hearings in the House, where the committee soon reported out a bill essentially unchanged from the War Department version. Elsewhere, however, the lobbying and public education efforts were better received. By the end of October, Truman was known to have reservations about the bill; he had endorsed it originally in the belief that it was uncontroversial. [78; 99]

The impasse in the Senate was resolved by the creation of a Special Committee on Atomic Energy. Grudgingly following its tradition, the Senate named freshman Brien McMahon of Connecticut, who had been first to suggest a new committee, as its chairman. As if to keep him in his place and to ensure that the committee would do nothing radical, it was filled with conservative Senate warhorses. During November and December, McMahon's committee learned as much as it could about the bomb and its implications, assisted by the new director of the National Bureau of Standards (a position some equated with the unofficial rank of the government's top physicist), Edward Condon. Despairing of nursing May-Johnson into acceptable health, the committee drafted an alternative bill. [78; 99]

The McMahon bill provided for a five-member full-time commission and a government monopoly on the production of fissionable material, and it prohibited active-duty military officers from serving as employees. Research with nondangerous amounts of fissionable materials would be without restrictions, there would be a policy that favored dissemination of information, and employees could not summarily be dismissed. These provisions ran counter to the actuality and intent of May-Johnson. Public testimony taken during four weeks of hearings early in 1946 altered the bill relatively little. Truman gave his endorsement, and after some further tinkering the bill passed the Senate in June 1946. When considered by the House, so many changes were made that the product resembled May-Johnson more than it did McMahon. In the conference committee of the two chambers the important provisions of the Senate version were retained. A good example is the question of patents. The Senate wanted AEC ownership of patents, while the House insisted upon private title. Indeed, government control of patents was labeled the "Russian" system, and the end of free enterprise was forecast. A rumor was circulated that steel had been found to be radioactive (it is, actually, to a small degree, as is almost everything) and that under the bill the steel industry would have to be nationalized. Despite these fears of creeping socialism, and fears generated by the revelation of a spy ring in Canada that had penetrated the Manhattan Project, leading to more calls for military control of atomic energy (illogically, for the espionage occurred under army supervision), the Senate version prevailed. On August 1, Truman signed into law the Atomic Energy Act of 1946. [78; 99]

Aside from the real questions of policy that were largely submerged under the rhetoric of civilian versus military control of atomic energy, there were troubling dilemmas raised by the scientists' movement itself. What was the proper role of scientific advisors? Should their contributions be limited to technical assessment, or should their value judgment also be sought? If their merit as advisors rests upon their technical skill, how can they maintain this competence if they spend much time away from the laboratory on public affairs? In fields burdened by secrecy, such that information is revealed to a limited number of people, how can the scientific community express a consensus? Finally, if scientists disagree on a question, is the scientific process flawed? [99]

In time, most of these problems resolved themselves. Various advisory committees were established by numerous governmental agencies, with presumably the appropriate ranges of expertise. Scientists were accorded the privileges of citizenship as much as lawyers and businessmen, and their opinions were given similar weight—which means they were listened to sometimes and ignored at other times. But it was recognized that their technical attention to topics of national interest inevitably led to political consideration of these same topics and that their insights should be valued. This involved greater sophistication on the part of the public, who gradually came to understand that scientists usually agreed on the technical details of an issue but could diverge on their interpretations of the data and disagree heatedly on any policy recommendations. Scientists were found to be capable of advisory roles, even if these took them from the laboratory, for, like other bureaucrats, they knew who to ask for the necessary information. In cases, they even violated the belief held by many that scientists should be "on tap, but not on top" by accepting appointments to important managerial positions.

If the scientists were initially uncertain of their role, the military were not. Defeat in retaining control of the nuclear weapons complex did not translate into hostility toward science. Quite the opposite, in fact. The military recognized that science would have a continuing major role to play in developing military hardware and in such areas as operations research; some forward-looking officers probably saw its use in strategic planning as well. At a time when the Atomic Energy Commission was being created and the National Science Foundation was a political football yet unborn, the military were science's greatest benefactors. The military had not only the funds to dispense but the wisdom to support basic research. They understood the scientists' argument that applied research and technology only reaped the fruits of fundamental studies, so they willingly created a population of scientists who worked in areas of unlikely military value but who felt comfortable being supported by the armed services. The relationship, albeit a controversial one, has been profitable for both sides. Despite some changes over the years, the Department of Defense still supports at least as much nonclassified research and development as do all the government's civilian agencies.

International Control of Nuclear Energy

Concurrent with the efforts for domestic legislation, the United States moved toward some form of international control. In October 1945, the president promised Congress that he would discuss with other nations means to share the benefits of atomic power and avoid the dangers of rivalry. He acknowledged that scientific secrets could not be kept but affirmed that he had every intention of withholding weapons technology from others. The United States regarded nuclear weapons as a "sacred trust" in its hands. Truman, however, delegated responsibility for international discussions to his secretary of state, and Byrnes was inclined to procrastinate. Tired of endless discussions on a range of topics at a meeting of foreign ministers, the image of Molotov asking to confer about the bomb gave him nightmares. But British prime minister Clement Attlee, buffeted between those who wished to use nuclear information to buy Soviet goodwill and those who scoffed at the idea that Stalin's behavior could be changed, insisted that some progress be made. [52:455–59]

This resulted in a conference with Truman in mid-November in Washington, where the details of an Anglo-American position were enumerated. The president and the prime ministers of Britain and Canada called upon the United Nations to establish a commission that would lay down the rules for the exchange of scientific information for peaceful purposes, control research and development to ensure only peaceful applications, devise the means to remove weapons of mass destruction from all nations, and establish safeguards to protect nations against violators of the new prohibitions. The UN's work would proceed step by step, the satisfactory achievement of one stage necessary before attempting the next. Such a process would allow the West to note the extent of Soviet cooperation, while minimizing its own exposure to danger. In January 1946, the UN created an Atomic Energy Commission (UNAEC), consisting of all the nations on the Security Council and Canada. [52:459–77]

To prepare the specifics of the U.S. position, Undersecretary of State Dean Acheson created a committee chaired by David Lilienthal, whose experience as head of the Tennessee Valley Authority gave him a rare insight into large-scale, public, technological operations. Panel member Oppenheimer spoke in favor of an international agency that would have research and development functions, both to attract top scientists and to be at the forefront of knowledge. The agency should have a monopoly on raw materials but should deal only with dangerous activities, such as the enrichment of U-235 or the operation of reactors in which significant amounts of plutonium might be produced. The emphasis was on positive contributions that would be made, not negative restrictions. [52:531–40]

By mid-March 1946, the committee, which largely agreed with Oppenheimer, presented its report to the secretary of state. At the same time, the president announced his appointment of Bernard Baruch as the American delegate to the UNAEC. The well-known 76-year-old financier was considered by many as an elder statesman, and his selection was seen as a measure of the importance the country gave to his task. But Baruch had no intention of being merely a messenger boy, carrying the so-called Acheson-Lilienthal Report to the UN; he insisted on a policy role. The key additions on which he insisted were that there be explicit penalties for violations of a treaty, and that these penalties be exempt from veto in the Security Council, where the Soviet Union was a permanent member. The State Department's view was that it was fruitless to enumerate penalties. If a major nation violated the treaty, the UN would be powerless to act, and war would be the only real corrective. It would thus be better to seek a treaty in the least intimidating atmosphere possible, for this would enhance the likelihood of its working. Baruch, however, was adamant and ultimately brought Truman to his side. [52:540–76]

The UNAEC opened its session in mid-June 1946, in New York City. "We are here to make a choice between the quick and the dead," Baruch began. The United States, he continued, wished that atomic energy be used for peaceful purposes, not for war. To achieve this, an international authority should be created to operate or own all activities that could be hazardous to other nations. But establishment of police provisions, not a research and development agency, was Baruch's main focus. If the United States was to give up its nuclear weapon monopoly, there must be "a program not composed merely of pious thoughts but of enforceable sanctions—an international law with teeth in it." This meant that a control system must be fully functional before the United States would cease to make its bombs and destroy its stockpile. Baruch itemized what he considered were some specific violations—going beyond his instructions from the administration—and asserted that punishments were central to the entire system. In this connection, he insisted, "there must be no veto to protect those who violate their solemn agreements not to develop or use atomic energy for destructive purposes." [52:576–79]

Two weeks after Baruch presented this plan to the UN, the United States began Operation Crossroads, test explosions of atomic bombs at Bikini Atoll in the Pacific, primarily to examine their effects upon naval forces. When the tests were first announced many months before, a number of scientists and senators protested that the United States would be accused of conducting atomic diplomacy: detonating weapons while simultaneously negotiating international agreements. Their prediction came to pass; the Soviets did challenge the honesty of American diplomacy. *Pravda* asserted that the United States had greater interest in perfecting the bomb as a threat against others than in eliminating it. Neither superpower chose to learn this lesson, however, for in the following decades both played the game in which Foreign Ministry or State Department appeared to be uninformed of military activities. In some cases this may have been true; in others the coincidence may have been coordinated to kill a diplomatic initiative while appearing to support it, and in yet others it is likely that one agency of government, in domestic political infighting, deliberately sought to sabotage efforts of another. [52:580–82]

Reaction to the Baruch Plan was largely favorable, though some saw it as giving away the U.S. advantage, while others regarded it as a proposal designed to be rejected by the Soviets, thus retaining the American monopoly. The Soviet representative, Andrei Gromyko, soon presented his country's position. First, there should be a covenant forbidding production, storage, and use of nuclear weapons. Then, within three months of the effective date of the pact, all existing weapons would be destroyed, and within six months nations would adopt strict punishments for violators. While he did not refer to the U.S. proposal in his speech—with one exception—Gromyko reversed its sequence by putting controls *after* disarmament. The exception was a clear statement that the Soviet Union would tolerate no changes in the veto power of the Security Council's permanent members. [52:582–85]

Further negotiations played more to public opinion than to serious compromise. Indeed, neither Baruch nor Gromyko was inclined to offer significant concessions. A Scientific and Technical Subcommittee formed in September 1946 was able to make progress on some issues, but knowing what was feasible mattered little without the political will to accomplish it. After more than 200 meetings the UNAEC ceased its efforts in the spring of 1948. [52:585–619]

Mutual distrust prevented an accord, as it would in so many future arms control arenas. The Soviets feared domination by the United States and its allies, while the West pondered whether Soviet intransigence was designed to buy them time to perfect their own nuclear weapons. And through it all there were remarkably few concessions to the anxieties of the other. A rigidity in attitude that dominated the Cold War fell quickly into place.

Cold War

In the twilight of his career, the distinguished American diplomat and historian George F. Kennan reflected upon the circumstances of the Cold War:

> When one looks at this relationship from the historical perspective, what one sees are two great powers only recently elevated to positions of political and economic ascendancy. One sees these two powers just beginning, in the 1930s and early 1940s, to tackle the difficult but not impossible task of psychological and political adjustment to each other in a world where new technology was making all men neighbors. But then one sees them suddenly overtaken by tremendous new developments in the geopolitical and military fields, developments for which they were not at all prepared; and one sees them thrown by these developments into a predicament—namely, the nuclear weapons race—that had nothing to do with those normal problems of adjusting to each other as they presented themselves in the 1930s, a predicament from which, as of today [1983], they know no means of escape, and in which they are simply writhing helplessly, at immense danger to themselves and to the world. [67]

The frustrating inability to deal amicably with each other that Kennan described colored the superpowers' postwar attitudes. The Soviets entered this period still stunned by the Nazi violation of their nonaggression pact and the devastating penetration of their borders. They had long memories of numerous other invasions of Russia by neighbors to east and west, and by 10,000 American troops (plus more from Britain, France, and Japan) during their civil war in 1918. They mourned the loss of over 20 million citizens in the recent war, and they believed that American lend-lease assistance was doled out during World War II at a rate cynically designed to help the Soviet Union defeat the Germans but to leave them nearly prostrate at the conflict's end. The Soviets clearly wanted buffer states around them, hence the establishment of satellite nations on their borders. They were suspicious of the goals of the United States and its friends, hence their intransigent behavior in the international arena. They were defensive of their national sovereignty, hence their resolve to retain veto power in the UN Security Council.

The United States, for its part, had held an unnatural terror of communism from its birth, which was not at all lessened by Soviet claims that it would destroy capitalism. Americans were almost panicked by the thought that communism, in the postwar chaos, would spread through both developed and developing nations and, especially, infiltrate into the States. Additionally, the United States now feared the military might of the Soviet Union and was uncertain how to bring democracy to the nations absorbed into the Soviet orbit. After World War I, newspaper columnist Walter Lippmann wrote that "the people are shivering in their boots over Bolshevism. They are far more afraid of Lenin than they ever were of the [German] Kaiser. We seem to be the most frightened lot of victors that the world ever saw." [72:A6] The same was true after World War II.

Many American politicians tied their careers to rooting out alleged subversives and enhancing national security. Senator Joseph McCarthy, who gave his name to an era of intolerance, was only the most prominent of them. Some historians suggest that the United States overreacted to the "Red menace," citing the small and weak domestic Communist party, the few spies ever actually caught, and the circumstance that the United States often ran the arms race against itself (building weapons in light of what Americans could engineer, instead of what the Soviets actually had). But the fear, rational or not, was real.

Even before World War II ended, Soviet attempts to establish puppet governments in Poland and Romania began to erode the goodwill most Americans felt toward them. As other Eastern and Central European nations also became satellites of the Soviet Union, and Communist parties in Italy and France showed surprising strength, the United States increasingly despaired being able to influence the Soviet Union with anything short of force. Former prime minister Winston Churchill, on a visit to the United States, dramatized the geopolitical split. In a speech on March 5, 1946, at Fulton, Missouri, he intoned the famous words: "From Stettin in the Baltic to Trieste in the Adriatic, an iron curtain has descended across the Continent." [89; 104:33; 119]

A year later, the president announced plans to aid countries to resist Communist pressures; this became known as the Truman Doctrine, of which Greece and Turkey were the principal beneficiaries. To counter Soviet expansionism, a policy of "containment"—effectively encircling that empire with military bases—was adopted. Western Europe, in economic turmoil and vulnerable in some countries to Communist party rule through free elections, was sent enormous amounts of American assistance. This was offered under the plan proposed in June 1947 by the former army chief of staff and then secretary of state, George Marshall. Foreign aid benefited the United States by creating overseas markets and enhancing national security. It was also much cheaper than fighting a hot war. But this remarkable act of international generosity was selfless, too, because there was no obligation to send the aid, and it was given in a strong humanitarian spirit. [89; 104; 119]

East and West became ever more polarized. Regional mutual defense alliances were formed, such as the North Atlantic Treaty Organization (NATO) and its Soviet-led counterpart, the Warsaw Pact. Within the span of a few years after World War II, American foreign policy made a remarkable break with its prewar paradigm. The Monroe Doctrine demanded that European nations refrain from inserting themselves in problems within the Western hemisphere, yet the United States, under the Truman Doctrine, now crossed that same barrier ocean to support one side in the Greek civil war. Similarly, the principle of nonintervention fell with Marshall Plan assistance to European countries, while creation of NATO ended the antipathy to peacetime alliances. In addition, the Cold War brought broadcasts on Voice of America and Radio Free Europe, peacetime conscription, the search for domestic subversives, the emergence of "Red" China in 1949, large military budgets, and the creation of what scholar Seymour Melman called the "permanent war economy," which was sponsored by what President Eisenhower termed the "military-industrial complex." [89; 104; 119]

With oscillations, these attitudes and practices continued for nearly half a century. While short on a historical scale, the period seemed endless with fear of the ultimate holocaust and was a bottomless hole for incredible amounts of national treasure. With the Cold War's end around 1990, as Eastern Europe broke away from the Soviet Union, and that empire itself dissolved in economic and geopolitical disarray, the West found itself faced with the need to define what it wanted in the "new world order." That process is still underway.

Atomic Energy Commission

Nuclear weapons formed the foundation of American resistance to the perceived Communist military threat. The source of these weapons was the Atomic Energy Commission, an agency rare in the federal bureaucracy for its strong technical orientation. Of the five commissioners appointed by Truman and approved by the Senate, one was a scientist, a tradition continued through the AEC's history.

In the year and a half between the war's end and the assumption of control by the AEC of its far-flung "empire" on January 1, 1947, the Manhattan Engineer District under General Groves continued to operate the complex. Without an unlimited budget, without a clear mandate to move in certain directions, and with unaccustomed public scrutiny, Groves nonetheless found it necessary to make some decisions.

The liquid thermal diffusion and electromagnetic separation plants at Oak Ridge were shut down, as the gaseous diffusion plant (and a sibling built nearby) was able to produce weapons-grade U-235 by itself. An experimental breeder reactor—one that created more fissionable material than it burned, by "breeding" plutonium from natural uranium—would be built near Chicago. Stanford physicist Norris Bradbury was named director of the Los Alamos Laboratory, replacing Oppenheimer, and living accommodations on the mesa and the water supply were improved. General Electric Company, which was interested in entering the commercial reactor business, succeeded Du Pont in running the Hanford facility. [52:620–33]

More interesting than these "housekeeping" actions was Groves's inclination and ability to act as a statesman of science. While his relationship with scientists was never very warm, he recognized that their research was essential for the future strength of the nation. But without stimulating opportunities for their investigations, they would have little desire to participate in weapons programs. In early 1946, he urged Army Chief of Staff Dwight Eisenhower to fund these activities in the next budget. Also, Groves appointed a committee of scientists which recommended creation of national laboratories, where unclassified basic research too expensive for universities would be performed. This led to formation of the Argonne National Laboratory, spun off from the Chicago Metallurgical Laboratory, the Brookhaven National Laboratory (on Long Island, New York), the Los Alamos National Laboratory, and several others. A pattern was set for a university or a consortium of universities to manage these facilities as AEC (and later Department of Energy, or DOE) contractors. [52:633–37]

The AEC commissioners, chaired by the former Tennessee Valley Authority head David Lilienthal, inherited from the army 37 installations in 19 states and Canada, worth more than the $2.2 billion spent during the war. While the commission began operation without an office and with no headquarters employees, there were 45,000 people, including those working for contractors, already on the job. The AEC decided to continue the system of hiring companies and universities to manage its facilities, instead of making all the staff AEC employees. In time, as civilian applications of nuclear energy increased, the commission created a regulatory staff to license such activities, in addition to its primary operating staff. By the 1970s, and after much criticism, the dual function of encouragement and regulation of an industry by the AEC led to establishment of the separate Nuclear Regulatory Commission (NRC). [5; 52:638–55]

Cold War pressures forced the AEC to devote almost all its early attention to weapons development and production. The first explosions in 1945 utilized only an estimated one-tenth of 1 percent of the fissionable material, so there clearly was a desire to design more efficient bombs. In a period when veiled threats of nuclear war against the Soviets were not uncommon, the size of the stockpile was also of great concern. The number of bombs in the arsenal was so secret that it was not committed to paper; an offical gave this information to the president orally. Decades later, when the size of the current arsenal customarily appeared in the newspapers, these ancient numbers were still denied to historians. The explanation was that the Soviets might be able to deduce from them the production capacity of the United States. Finally, in 1982, they were released. The number of bombs (all of the implosion type) was surprisingly small: 2 in 1945; 9 in 1946; 13 in 1947; 50 in 1948. [57] This shows not only that the early weapons were handcrafted, not made on an assembly line, but also the extent to which the Manhattan Project factories and laboratories were weakened by the exodus of personnel after the war and the uncertainties created by the legislative battle over creation of the AEC.

Once the AEC got organized, it contracted to purchase uranium ore from South Africa, the Congo, Canada, Australia, and, mostly, from domestic miners. By the late 1950s, these efforts were so successful that there was an oversupply, and the AEC terminated some contracts and stretched out others. When domestic utilities began to build reactors for the production of electricity in the 1960s and 1970s, mining activity increased once again. [5]

Many new facilities were constructed in the 1950s. Concentrate from the ore mills, containing 50 to 80 percent uranium, was shipped to "feed materials" plants at Fernald, Ohio (operated by the National Lead Company of Ohio), and Weldon Springs, Missouri (run by the Mallinckrodt Chemical Works), where high-purity uranium trioxide was produced. This, in turn, was converted into uranium hexafluoride for the gaseous diffusion process or metal for reactor fuel elements. In addition to the facilities still operating at Oak Ridge, Tennessee (managed by Union Carbide Company), the AEC built other gaseous diffusion plants at Paducah, Kentucky (Union Carbide was the contractor here, too), and Portsmouth, Ohio (operated by Goodyear Atomic Corporation). The Hanford, Washington, site (where General Electric Company and Battelle Memorial Institute were cocontractors), became home to a total of nine plutonium production reactors, while five more were built at Savannah River, South Carolina (run by Du Pont). [5]

In 1953, President Eisenhower announced his Atoms for Peace program, in which the United States would help other nations to gain the beneficial uses of nuclear energy in medicine, food preservation, industrial testing, electricity production, and other applications. At home, electric utilities were encouraged to purchase reactors, from such companies as General Electric and Westinghouse. Nuclear-generated electricity, in a phrase attributed to AEC chairman Lewis Strauss, was expected to be "too cheap to meter," meaning it would cost more to send homeowners a monthly bill than to give the electricity away free. Of course, that never happened. Poor construction planning and execution, primarily, but also public fear of accidents and opposition by public interest groups stretched out the licensing and construction timetable so much that not only was no electricity given away free but many utilities, to stem the hemorrhage of expenses, canceled the reactors they had ordered in the optimistic 1960s and early 1970s.

The AEC sponsored another program worthy of note, called Project Plowshare. The name was chosen from the biblical injunction to beat swords into plowshares, and the intent was to show that nuclear explosives could be used for peaceful purposes. Placed deep underground, an explosive device could fracture and heat oil-bearing shale, thereby facilitating the extraction of oil and gas. The same principle might work with the recovery of a water-soluble mineral, such as sulfur. If the nuclear charge was placed in a shallow location, it might excavate a harbor, as was contemplated for Alaska's north coast, or (using literally hundreds of explosives) carve out a sea-level canal connecting Atlantic and Pacific, as was proposed for Colombia and Nicaragua. As it happened, a series of test explosions throughout the 1960s yielded disappointing results. The concept failed on the basis of cost effectiveness and the problem of radioactivity (escaping into the atmosphere or in the extracted mineral), and Plowshare rusted to a close in the early 1970s. [5]

From the scientists' perspective, the AEC's most worthwhile activity was its support of unclassified basic and applied research. Physics, chemistry, biology, medicine, environmental studies, agriculture, mathematics, computer science, geology, metallurgy, and other fields enjoyed generous support. Work was conducted in universities across the country and at specially funded facilities, such as the particle accelerator laboratories at Brookhaven National Laboratory, University of California at Berkeley, California Institute of Technology, Harvard, Princeton–University of Pennsylvania, Stanford, and Argonne National Laboratory. This reliance upon government funding illustrated the circumstance that no university, industry, or foundation was wealthy enough to support science alone or even collectively. Indeed, the AEC (now DOE), National Science Foundation, National Institutes of Health, and Department of Defense have funded the lion's share of science in the United States since 1945. This may be regarded as proper recognition of a national resource, but it also injected into science a bureaucracy that sometimes wishes to call the piper's tune. [53:1–361]

The Monopoly Broken

President Truman created widespread shock with his announcement on September 23, 1949, that one month earlier the Soviet Union had exploded a nuclear device. High-flying aircraft, specially equipped with filters, had picked up unusual amounts of radioactivity above the north Pacific Ocean, and analysis had shown the radioactive particles to be fission fragments. Many could not believe that the Soviets were capable of such a "high-tech" feat; at worst it must have been an experimental reactor that blew up. But if it really was a bomb, they reasoned, its construction must have been aided by spying on the American project, for the Russians were regarded as bumbling peasants, incapable of sophisticated research and development. Further, any bomb they might have was surely inferior to the new and improved models that the United States had tested at Eniwetok Atoll. America need not despair after all, since the Soviet Union lacked the industrial infrastructure needed for nuclear weapon production, as well as long-range bombers needed for delivery. [47:137–43; 53:362–75]

The Franck Report to the secretary of war in June 1945 had warned that any industrialized nation determined to build nuclear weapons could repeat the U.S. success in three or four years, an estimate echoed by other scientists. Most people, however, seem to have preferred the more reassuring testimony that General Groves gave to Congress in November 1945, when he said it would take the Soviet Union 15 to 20 years to duplicate the American effort. [47:22, 135–43]

Despite this background and the efforts to minimize the Soviet achievement, the news came as a jolt. American politicians, scientists, and opinion leaders sought an appropriate response. This was an effort to make thermonuclear weapons (hydrogen bombs, to be described in the next chapter). That a high-tech, hardware response was chosen illustrates the low status of arms control expectations as well as the cyclical, or action-reaction, nature of the escalating arms race.

American scientists who knew their Soviet counterparts were not overly surprised that they produced a bomb in 1949. Physicists such as Peter Kapitza, George Gamow (who emigrated to the United States in the 1930s), and Lev Landau were impressive products of the Soviet educational system. These were men familiar on the international scene, and, indeed, Kapitza and Gamow had received some notoriety in the 1930s, when their government sought to employ their skills against their wishes. At that time, in fact, the Kremlin benefited from a wave of sympathy, for it wished to retain their services at a time when Nazi Germany was expelling so many of its well-trained citizens. [12; 63]

Nuclear physics had flourished for some years in Leningrad, in Moscow, and at a new center in Kharkov, and Soviet physicists, like their colleagues elsewhere, rushed to examine fission when the phenomenon was announced in early 1939. Georgi Flerov and Konstantin Petrzhak published in the American *Physical Review* their discovery that uranium may fission spontaneously, not just when humans place the element in a neutron-rich environment. Yakov Frenkel, simultaneously and independently of Niels Bohr and John Wheeler, produced a theory of fission based on a novel model of the nucleus. These were world-class contributions to nuclear physics and served as evidence of that nation's capabilities. One must not judge all of Soviet science by Trofim Lysenko's mindless destruction of genetics. [45; 86]

Fission was discussed openly in the Soviet Union until the time of a physics conference in Moscow in November 1940, at which the participants debated the need for the government to fund construction of a reactor. The possibility of nuclear weapons, obvious to most scientists, was brought to the attention of the People's Commissariat of Heavy Industry. These early efforts were interrupted in June 1941, however, when their former ally, Germany, invaded Soviet territory. The contents of laboratories and factories were moved eastward for safety, and scientists turned their attention to more pressing military needs, such as radar and means to protect ships from mines. [45; 58]

When the tide of battle turned and Soviet forces halted the advance of German troops, in the summer of 1942, nuclear physics reclaimed the interest of some scientists. It is likely that the government encouraged this upon learning that both the United States and Germany were secretly investigating the potential of nuclear explosives. By the end of the year, Igor Kurchatov was appointed scientific director of atomic research. With authority to recruit widely, he commandeered scientists from military units, industrial laboratories, and research institutes and moved his staff of about 50 people into quarters that remained empty from the evacuation of Moscow. [45; 58]

The scale of the effort was much smaller than that in the United States, but it is still remarkable that so much was accomplished in a country fighting under such hardships. They ordered uranium and graphite of high purity, for a pile to be built on the same principle as Fermi's in Chicago; heavy water (containing hydrogen's heavier isotope, deuterium) was seen as an alternative neutron moderator. They rushed to complete construction of a few cyclotrons, and they measured a variety of nuclear constants. Research expanded as the Soviets won the gigantic battle of Stalingrad in February 1943 and raised the siege of Leningrad in January 1944. As was the case in the Manhattan Project, Kurchatov pursued a variety of techniques toward each end, until a technique was proven unusable and dropped, or one was fully successful and the others could be abandoned. [45; 58]

News of Hiroshima and Nagasaki caused them to redouble their efforts, for now they knew that their goal was attainable. Industrial development expanded greatly, and experienced managers were drawn from the wartime factories and bureaucracy. Several methods were investigated for separating uranium isotopes: electromagnetic separation, gaseous diffusion, and thermal diffusion. During the next year the quality and quantity of reactor materials improved, and in December 1946, just four years after Fermi's success, Kurchatov initiated the first controlled chain reaction in Europe. The reactor was fueled with natural uranium and moderated by graphite. Next, Kurchatov, who played something of the combined roles of Vannevar Bush, Arthur Compton, Enrico Fermi, and J. Robert Oppenheimer, gleaned data from this experimental pile to design a production reactor. Construction began in late 1947, and in time plutonium was delivered to the weapons laboratory. The Soviet equivalent of the Trinity test, using plutonium, occurred in August 1949. The American penchant for nicknames—it was widely called Joe-1, after Joseph Stalin—belied the dread that news of the Soviet bomb occasioned. [45; 58]

That the Soviet bomb benefited considerably from their spying on the Allied project—they copied many procedures and designs—is now admitted. But this detracts only a little from their high level of scientific and industrial accomplishment.

Living with the Cold War

The Thermonuclear Decision

HOW BEST TO respond to the Soviet acquisition of nuclear weapons? Conciliation has generally been regarded as a weaker reaction to a potential threat than an arms buildup. Nevertheless, a few voices suggested that this balancing of power might be a suitable cloak for the pinstriped diplomats to pursue arms control and generally improve superpower relations. But the stiff breezes of the Cold War tore such fabric to shreds. Both superpowers and their allies settled into an uncomfortable, expensive, all-absorbing, and occasionally terrifying state of hostility. Although they came close to war in the Berlin blockade of 1948 and the Cuban Missile Crisis of 1962, the United States and the Soviet Union never faced each other across a battlefield; hence the name "Cold War." Yet each engaged in armed conflicts against a proxy of its real opponent, as in Afghanistan, Korea, and Vietnam, and each supported countless insurgents against clients of the opponent, as in Angola, Ethiopia, Somalia, Cuba, Iran, Nicaragua, and Guatemala.

If one side scored a triumph, the other felt compelled to respond. International prestige was no less important than territory and resources. Nowhere was the competition more apparent than in the hardware of warfare. The United States usually led in this race, but the Soviets inevitably matched the accomplishment. Thus was the arms race ratcheted ever upward. Thermonuclear weapons were an important step in this escalation, providing vastly more explosive force than fission devices of the same weight. In time, they came to dominate the arsenals of both nations, not only in multimegaton bombs but also in low-yield tactical weapons.

Shortly after the Hahn-Strassmann discovery that uranium fissions when bombarded by neutrons, Cornell University physicist Hans Bethe,

80

in unrelated work, published a paper on energy production in stars. It had long been understood that ordinary chemical combustion could not be the mechanism that kept stars burning for billions of years; the temperatures were too high and the fuel reserves too low. Physicists recognized that nuclear reactions must be involved. Bethe's contribution was the explanation of the carbon cycle, in which lighter elements join together to form carbon and then decompose to allow the process to repeat. Progress in science depends heavily on quantification, and Bethe provided calculations of reaction rates and energy production. On earth, nuclear reactions are not routinely apparent to most people; we find them in natural radioactivity, particle accelerating machines, and reactors (not to mention bombs). In stars, however, these reactions are central to existence. [113]

The energy is released when nuclei of light atoms join, or fuse, together; this is the opposite of the fission process, wherein nuclei of heavy atoms split apart. Clearly, the most stable atoms are those of middle size. When the discovery of fission galvanized physicists' thoughts upon weapons, attention was also given to fusion. Thermonuclear (fusion) bombs, using isotopes of hydrogen, were potentially easier and cheaper to construct than uranium or plutonium explosives. Hydrogen was abundant—in water, for example—and its heavy isotope, deuterium, has a mass double that of the most common hydrogen, making its separation far easier than peeling away U-235 from U-238.

Fusion had, in fact, been accomplished in Rutherford's Cavendish Laboratory in the mid-1930s, but far more energy had to be put into the system to cause the reaction than could be extracted. If, however, a temperature comparable to that found in the interior of stars could be reached, unlimited quantities of hydrogen fuel would fuse. Indeed, fusion was more efficient than fission. In fusion, all the reacting materials are consumed, and a bomb of any size can theoretically be constructed; whereas in fission, because the pieces of heavy elements are blown apart, there is a maximum size of about one megaton. The hurdle, then, was to achieve a temperature found nowhere on earth.

When Oppenheimer organized a meeting of theoretical physicists at Berkeley in the summer of 1942, fusion was discussed. No one was more eager to explore this reaction than Edward Teller, the young Hungarian refugee who had helped Szilard orchestrate Einstein's letter to President Roosevelt. But fission was the only mechanism seen to create the fusion temperature, so priority naturally went to fission investigations. One offshoot of the fusion inquiry, however, was the consideration of whether a fission bomb or a fusion bomb would kindle hydrogen or nitrogen or some other component of the atmosphere and literally ignite the globe.

This was a problem given the serious attention it deserved on more than one occasion, with the conclusion that it was impossible.

At Los Alamos, fusion investigations led by Teller received little support. Even after the successful Trinity test of the plutonium bomb, the way was not clear to hydrogen weapons—it requires more than just placing a mass of hydrogen next to a fission bomb—and with the exodus of personnel after the war, fusion studies were dropped completely in June 1946. Teller departed, too, and from the University of Chicago continued to advocate an all-out effort for the superbomb. When Joe-1 was detected, he was looked upon as a prophet. Thermonuclear weapons would lead the United States back to the promised (if ephemeral) land of "superiority." [52:240, 631–32]

The AEC's initial response to the Soviet bomb was to increase production facilities for fission weapons. Others in government looked to civil defense as an answer. But energetic lobbying by Teller and by Berkeley physicists Ernest Lawrence and Luis Alvarez, made it clear that "the Super" must also be considered. The congressional Joint Committee on Atomic Energy was enthusiastic about the idea and, on six separate occasions between September 1949 and January 1950, urged President Truman to initiate a crash program for thermonuclear weapons. California Congressman Chet Holifield, a committee member, translated a 20-megaton bomb (over 1,000 times the yield of the Hiroshima weapon) into "everyday" terms; it would be, he said, the equivalent of a train of boxcars loaded with TNT, extending from Boston to Los Angeles. [53:362–80; 61:45]

Since its creation, the AEC had relied for counsel upon a panel of high-powered scientific advisors called the General Advisory Committee (GAC). This body recommended a modest program of fusion research, and beginning in 1947 this was conducted at Los Alamos. They hesitated to push for more, since no one knew if the problem had a technical solution; laws of nature might prevent construction of a hydrogen bomb. Now, needing a fitting response to Joe-1, the AEC commissioners asked the GAC to reconsider the Super. [53:380–85; 121]

This was really a political problem, for the U.S. arsenal, however small, was larger than that of the Soviets and could likely be maintained so. But the Truman administration (and its successors) never tired of flaunting its atomic monopoly or its leadership in some other category (usually numbers of weapons) and now was hard-pressed to reassure the public that the U.S. was "ahead" and therefore still "safe." The Soviet Union, for its part, felt that nuclear weapons were essential for its own security. Further, it was determined to be, and be globally recognized as, a superpower. Nuclear weapons were the vehicle for this success, even if the nation's economy had to be sacrificed.

The GAC met in Washington at the end of October 1949. The ubiquitous Oppenheimer, who seemed to lead so many important panels, or at least compose their reports, was chairman. After much analysis and soul-searching, he and his colleagues unanimously recommended to the AEC that it *not* proceed with the H-bomb. They argued that the large yield of such weapons was the reason for interest in them, but this enormous power was excessive for military installations. Cities, obviously, would be the targets, and the Super must therefore be viewed as a genocidal weapon. Vast quantities of radioactivity would contaminate huge areas, adding to the devastation. Should smaller thermonuclear weapons be possible, the uncertainties of design and development made it unclear if they would be any cheaper than fission bombs. [53:380–85; 121]

A GAC minority, consisting of Enrico Fermi and I. I. Rabi, added that "such a weapon cannot be justified on any ethical ground which gives a human being a certain individuality and dignity even if he happens to be a resident of an enemy country." They believed that "it is necessarily an evil thing considered in any light." The United States should lead other nations in a pledge not to make these weapons; if anyone violated the oath, the US stockpile of atomic bombs was adequate for retaliatory purposes, or the United States could then decide to make hydrogen bombs itself. [121:152–59]

The GAC majority, which included Oppenheimer, James B. Conant, and Caltech president Lee DuBridge, went even further and recommended a unilateral commitment never to construct thermonuclear weapons. They argued that the Super "is in a totally different category from an atomic bomb," and "its use would involve a decision to slaughter a vast number of civilians." They concluded: "In determining not to proceed to develop the super bomb, we see a unique opportunity of providing by example some limitations on the totality of war and thus limiting the fear and arousing the hopes of mankind." [121:152–59]

This report to the AEC followed in the tradition of the Metallurgical Laboratory's Franck Report of 1945. Originally, scientists rarely had the opportunity to offer political advice; the advent of nuclear weapons raised them to sufficient prominence that their words now were taken seriously. Not that this made them infallible oracles, or that the scientific community was ever thoroughly united on any issue. Nonscientists asked why these laboratory inhabitants thought they had adequate familiarity with the ways of the world to offer nontechnical advice. Scientists responded that they were speaking merely as citizens who had taken enough interest in a subject to develop some expertise.

They also defended their democratic right to present their opinions, and many saw themselves as part of a trend, continuing from the 1930s, of relating science to social responsibility. Nonetheless, whether scientists should be "on tap" only for their technical expertise, or "on top" to assist in policy formulation or even be bureaucrats, is still a heated issue for some.

The GAC report was controversial. True, the committee's mandate included advice on policy, but it was rare for recommendations to be couched in moral and ethical terms. The AEC commissioners split three to two in favor of the report. Their assorted views then went to the president the second week in November. Because Truman recognized that political and military considerations, in addition to technical factors, were involved in the Super decision, he referred the matter to a subcommittee of the National Security Council (NSC), composed of retiring AEC chairman David Lilienthal, Secretary of State Dean Acheson, and Secretary of Defense Louis Johnson. [53:388–91]

Lilienthal, long bothered by the tendency of the military to base their statements of their nuclear "needs" simply on how much the AEC could produce (often with a little more requested as an incentive for the AEC to work harder), urged a review of foreign policy and defense strategy before making a decision on the Super. Were nuclear weapons held only for deterrence and retaliation, or were they to be so completely integrated into the fighting forces that their use would be almost inevitable (as he believed already had happened)? Whatever the policy, the United States should formulate it consciously, not drift into it. If the former was the guideline, the stockpile of fission bombs was adequate, given the value the Soviets placed upon their easily targeted factories. This was an opportunity, Lilienthal thought, to step back from the tensions of the arms race by deferring pursuit of the Super, in order to seek better relations with the Soviet Union. [93]

To Johnson, the GAC and AEC's argument that fission weapons were all that the U.S. military needed was ludicrous. With the same logic that inspired the search for fission weapons a decade earlier, he contended that our troops would be at a severe disadvantage if only the enemy had hydrogen bombs. From Acheson's perspective, the Soviets would reap political gains if they acquired a monopoly on the Super. He further judged that they were not interested then in arms control talks. When the subcommittee voted, Defense and State united against the AEC. [93]

Truman received the NSC report in January 1950. On the last day of the month he announced that the United States would investigate the technical likelihood of producing thermonuclear weapons. At the same time, State and Defense would conduct a review of U.S. foreign and military policies. Lilienthal would get his analysis, though not before the H-bomb decision was made. Yet it was a curious decision, called by one author a choice that settled very little. The size and pace of the thermonuclear program were not specified. A fall in atomic bomb production (by diverting one or more reactors away from plutonium production) was not examined. The utility of strategic bombing with enormous weapons—indeed, any use of them—was not explored. The value of an arms agreement with the Soviet Union was not considered. [93]

Behind Truman's announcement lay geopolitical realities. As seen by the United States, the Soviet Union was an implacable adversary, intent on world domination. Friendly actions on its part were interpreted as efforts to lull the United States into a false sense of safety. Prior to World War II, American security was threatened only by the improbable coalition of several European powers, for small groups of nations could not match the industry, resources, and population of the United States. After the war, and despite the massive aid given to Western Europe, those countries failed to forge a new balance of power, leaving it to the United States to compensate for Soviet strength. Though allies were involved in NATO and in other pacts and programs, American might was the bedrock of efforts to resist Soviet pressures. [51; 93]

Once the Kremlin acquired nuclear weapons, American invulnerability ended. A nation so armed could quickly attack the United States, without the time-consuming (and hard to conceal) mobilization of troops within its borders. American allies abroad would no longer slow an enemy's advance, nor would the vast oceans continue to serve as buffers, giving the United States time to assemble its forces. Only American military strength could deter Soviet power. Only by holding a "winning weapon" could the United States prevail. [51; 93]

Political verities meshed with geopolitical realities; the American public had an abiding fear of the Soviet "menace" and expected the White House to respond forcefully to Joe-1. If additional pressure was needed to convince the president, some cases of atomic espionage were revealed, most notably the arrest of Klaus Fuchs in England as a Soviet spy in January 1950. Again, the public would expect some type of response.

Truman recognized these circumstances, and for him it was only logical that fusion must be investigated. This was, as mentioned above, a minimal decision, but it did get the "camel's nose under the tent"—sometimes a politically expedient maneuver. Within a month the Joint Chiefs of Staff pushed through the rest of the camel by requesting an all-out program, not just for exploring fusion but for developing hydrogen bombs, producing them in quantity, and acquiring means for their delivery. Truman soon approved this, ordering construction of new reactors at the Savannah River site in South Carolina. [93]

Constructing Hydrogen Bombs

At Los Alamos the theoretical division doubled in size. A number of bright young men joined the program, and several prominent physicists, including George Gamow, Edward Teller, John Wheeler, Richard Garwin, John von Neumann, Enrico Fermi, and Hans Bethe, came for longer or shorter periods. The appearance of Fermi and Bethe may seem puzzling, as both were on record as opposed to the Super. Fermi's participation was explained as that of a loyal citizen who dropped his own protest once legal authority chose otherwise. Bethe seems to have decided to cooperate in hopes of proving that thermonuclear weapons were physically impossible to make. The outbreak of the Korean War in June 1950 resolved doubts held in a number of minds. [121]

Bethe had first declined Teller's invitation, leading the latter to suspect that Oppenheimer had influenced him. This was one of the points of simmering hostility between Teller and Oppenheimer, which (as will be seen below) reached ignition temperature in 1953. Oppenheimer claimed that he ceased to oppose the Super once Truman announced his decision, and that Teller exaggerated his influence over others.

The Los Alamos team devised an experiment of the boosted fission or "booster" principle, in which a large fission explosion ignites a small amount of thermonuclear fuel, the object being to increase the yield of uranium and plutonium weapons. They also conceived of a preliminary step toward the more important arrangement, a superweapon, in which a relatively large mass of thermonuclear fuel is kindled by a small fission bomb. Both principles were successfully tested at Eniwetok Atoll in May 1951. [121]

Final breakthroughs for construction of the Super occurred the same year. Mathematician Stanislaus Ulam contrived a new configuration for the ignition of nuclear weapons, and Teller modified it to a more convenient and general form. Extensive calculations were made possible by newly invented electronic computers, such as UNIVAC and MANIAC. The new geometry involves an interior surface of the bomb casing that reflects X-rays and gamma rays from the fission trigger, allowing radiation pressure to compress and heat the deuterium fuel. [121]

Late in 1951, Teller returned to the University of Chicago. He had developed a promising concept, and others could bring it to fruition. Teller, further, had not been happy at Los Alamos, where he differed with director Norris Bradbury over the shape and direction of the fusion effort. As an outsider he was free to lobby in Washington for changes.

With support from the air force, which threatened to create a nuclear laboratory of its own, and from Ernest Lawrence at Berkeley, the AEC was pressured in 1952 to establish a second bomb laboratory. Its role was to provide competition to Los Alamos as well as alternative designs. Initially little more than an offshoot of Lawrence's famous Radiation Laboratory, where he built his cyclotrons, it was located in Livermore, California; one of Lawrence's young colleagues, Herbert York, became its first director. [121]

Los Alamos planned two distinct tests. The first was an experimental confirmation of the Ulam-Teller innovation. This was not a weapon, it was a physics laboratory: a huge device festooned with instruments, weighing about 65 tons, with pressure vessels and cryogenic apparatus to liquefy deuterium gas at temperatures below $-250°C$. Erected on Elugelab Island at Eniwetok, the "Mike" device was successfully detonated on November 1, 1952. The yield, 10 megatons, was far beyond any previous explosion caused by humans. The second test was of a military configuration of the Super; it had to fit into an aircraft. Not only were size, shape, and weight now of crucial importance, but the deuterium had to be fixed in a dry compound at normal temperatures. On March 1, 1954, the "Bravo" device exploded, with a yield of 15 megatons, depositing radioactive ash far downwind. [121]

Just as scientists the world over recognized the explosive potential of nuclear fission reactions, they also comprehended that fusion reactions would liberate much energy. It is no surprise then that the Soviets' thermonuclear efforts appear to have begun before Truman's announcement of the American program in January 1950. They had been informed of American discussions about thermonuclear weapons by Klaus Fuchs but could have benefited little technically because his contact with American developments ended long before the advent of a feasible concept. Primary impetus must be ascribed to the creativity of Soviet scientists, still led by Igor Kurchatov, but with the youthful Andrei Sakharov (who later won the Nobel Prize for Peace for his activities as a dissident) as the crucial theoretician. [45; 121]

Joe-4, the fourth Soviet nuclear explosion and the first to contain a thermonuclear burn, was detonated on August 12, 1953, just nine months after the American Mike test. Unlike Mike, it had a more advanced dry deuterium fuel. Again unlike Mike, it was a booster (a fission explosion enhanced by a small hydrogen burn), not a Super, having a yield probably below 400 kilotons. These distinctions were not explained publicly for some decades, leading to a variety of interpretations of who was "ahead" in the nuclear arms race. The Soviets' first Super, dropped from an airplane, was detonated on November 23, 1955, with a yield of a few megatons. [121]

American hydrogen bombs in the 1950s could only be delivered by aircraft. This allowed them to be of large size, as much as a few tens of megatons. With the introduction of intercontinental ballistic missiles (ICBMs) in the very late 1950s, smaller warheads were necessary to match the rockets' smaller carrying capacity, and the largest yields were about five megatons. As missile accuracy increased, warhead yield could be further reduced with the same "kill probability." Yield was reduced still more with the development of multiple independently targetable reentry vehicles (MIRVs) for the Minuteman ICBM (hundreds of kilotons) and the Poseidon submarine-launched ballistic missile (SLBM) (tens of kilotons), in which up to 14 warheads were placed atop a single rocket. Thus, while warheads were constructed in the kiloton range, a single one could destroy a medium-sized city, while a barrage from MIRVed missiles could level a large metropolitan area. Thermonuclear explosives of extremely low kiloton yield were also produced for use in a variety of tactical (battlefield) weapons first developed in the 1950s.

The Soviets followed essentially the same path, with at least one interesting difference. Americans delighted in sophisticated technology, microminiaturizing equipment wherever possible to reduce weight. Soviet designers followed more of a brute-force path. The Soviets built larger, more powerful rockets, which were also of value when they wished to loft a heavy payload into space. They also built larger H-bombs. The largest ever tested, in October 1961, was 58 megatons. It was hard to think of a target "worthy" of such size. London could be destroyed to a radius of ten miles with a 20-megaton weapon.

Geopolitical Problems

From 1945 until the end of the Cold War around 1990, every U.S. president spoke eloquently about the futility of the arms race. Why futile? Because in those 45 years the United States spent several trillion dollars on defense, yet the Pentagon could not prevent the annihilation of the nation. Was this expenditure then to no avail? A variety of responses is possible, none able to be proved correct.

The United States seemed trapped in this spending cycle, since the nation followed to a large extent the concept of "peace through strength." This can be phrased in many ways, but it is essentially part of the tradition of deterrence, which was not a new concept in the 1970s but rather a pattern of behavior dating to earliest times. Stones, clubs, and spears in the hands of one tribe surely made a potential opponent think twice before attacking. In the nuclear age, deterrence was defined to be the ability to inflict unacceptable damage upon an enemy—which would prevent the attack.

The United States and the Soviet Union were but the latest in a long series of antagonists who hoped to avoid war but armed to deter it, or to prevail if it came to blows. Most people believed that capitalism and communism were locked in mortal combat. Opponents of this mind-set suggested a number of alternatives. To reduce or remove the suspicion and fear that fanned the winds of war, they recommended that nations should engage in scientific and cultural exchanges, military commanders should inform each other of exercises that might be regarded as real mobilizations, government leaders should consult frequently via summit meetings and electronic communications, and commercial trade should be encouraged. In other words, the nations should endeavor to act in the ways of friendly states, in hopes that the means would eventually become indistinguishable from the ends. Another suggestion was the acquisition of military forces that were capable of defensive but not offensive action. And yet another involved relinquishing national sovereignty to a degree, allowing the United Nations to keep the peace. Scientists were active in proposing and supporting many of these measures.

All of these alternatives were actually tried, at least in part, and some were considered successful. But the yardstick of success was "how bad it might have been otherwise," and not whether the arms race was ended. Fear and suspicion were too strong, on both sides. Emotionally, logically, and politically, it was impossible to put the nation's survival at risk. Unilateral arms reductions, even though made by both superpowers in the course of restructuring their forces, were seen as foolish expenditures of diplomatic capital; it was more sensible to use reductions as "bargaining chips."

Fear and suspicion also pushed each nation into extreme positions. Robert McNamara, secretary of defense in the Kennedy and Johnson administrations, pointed out that a strategic planner had to design forces and responses for the worst possible case, not merely for the most likely. "Worst-case analyses" thus became a common ingredient in defense thinking. This highlights another dilemma of the arms race. The military was often criticized for its insatiable demand for more and better weapons as it planned for these worst cases, and for the enormous defense budgets thus required. Yet citizens expected the military to protect the nation in an emergency, with minimal loss of life and property. In the nuclear age, one felt that there would not be a second chance.

Rarely, however, was it argued successfully that force reductions might reduce tensions, or more fundamentally, that the crucial feature was not the number of weapons but the relationship between the superpowers. In the United States, where supporters of a large military establishment gathered to themselves the accoutrements of patriotism, especially the flag, advocates of improving international relations not by coercion but by peaceful self-interest were all too easily branded as "pinkos" or "commie sympathizers" and told they could leave the country if they did not like the "hawkish" atmosphere: "Love it or leave it."

There have been no world wars since 1945, but that is not to say the world has been at peace. At any given time, there are several conflicts raging in one part or another of the globe. Of particular concern to the superpowers was the Korean War (1950–53), in which the United States, under the banner of the UN, pushed back North Korean and Chinese troops above a line of truce across the peninsula. There were cries in the United States that America had invited this Communist aggression by placing Korea beyond its perimeter of defense. There were also calls for the United States to use nuclear weapons against its adversaries.

Communist leader Mao Zedong had liberated China from the nationalists in 1949 (with finger-pointing in the United States at those who had "lost China" to the free world—as if China were anyone's to lose), following his concept of a "people's war." But in Korea he found that his vast manpower was decimated by the Americans' greater firepower. This circumstance, and veiled threats by the United States to use nuclear weapons if the People's Republic of China (PRC, often called Red China and Mainland China) sought to invade Taiwan, where the nationalists took refuge, convinced Mao that the PRC needed nuclear bombs of its own. Until they had a falling-out, the Soviet Union assisted China in this effort; thereafter on its own, China detonated its U-235 weapon in 1964. [71]

In 1956, Hungarian efforts for political reform were met with such resistance by their Soviet overloads that the people rose up in rebellion. The revolution, which many believe was encouraged by implied promises of assistance from the West, was crushed by Soviet tanks and troops. In the same year, the armies of Israel, Great Britain, and France invaded Egypt, ostensibly to place in international hands the Suez Canal, which President Gamal Abdel Nasser had nationalized, but more likely to topple Nasser and remove a threat to Israel's safety. President Eisenhower brought pressure on all three allies to withdraw. More ominous, Premier Nikita Khrushchev insisted that the invasion stop or Britain and France might come under nuclear attack. This was a measure of the Soviet

Union's growing missile capability (intermediate-range in this case), while both Hungary and Suez mark occasions when the United States stepped back from confrontation in fear of global nuclear war.

Arms Race

While the United States and the Soviet Union were perfecting their hydrogen devices, Great Britain developed its own atomic bomb, which it tested at an Australian site in October 1952. Five years later, it too had thermonuclear weapons. France had nuclear weapons by 1960 and an H-bomb by 1968; India successfully tested a nuclear device built, its government said, only for peaceful purposes, in 1974. (This is not entirely preposterous, for, as described above, large underground explosions can be used for mineral extraction and earth moving, as was studied in the U.S. project called Plowshare.)

In following decades, a few nations, such as Israel and South Africa, were widely credited with nuclear arsenals; Israel never confirmed this capability, but in the early 1990s South Africa admitted to having had constructed and then dismantled a handful of nuclear weapons. Other countries, including Pakistan, Iraq, and Libya, were believed to be making serious efforts to join the "nuclear club." But at the time of this writing (1994), there are some hopeful signs that nuclear proliferation may be reduced, as wary neighbors, such as Argentina and Brazil, and North and South Korea, have proclaimed their abandonment of the "nuclear option." Yet, dangers still exist, as North Korea is alternately cooperative and evasive with the International Atomic Energy Agency's inspection team, which is concerned about the diversion of plutonium from spent reactor fuel, a situation that could impel South Korea and even Japan to enter the nuclear club. Recognizing the impossibility of totally preventing proliferation, the U.S. government is contemplating a new policy of attempting merely to limit the spread of nuclear weapons by banning the export of necessary technologies.

In the period covered by this book governments built large science/technology infrastructures, and scientists were involved in all these events not just for the obvious purpose of designing and constructing weapons. Science was necessary also to evaluate what other nations were doing (e.g., by gathering and analyzing satellite surveillance data) and to help negotiate various treaties (e.g., by developing verifiable conditions).

As the superpowers perfected new warheads in the 1950s, for gravity bombs, missiles, and a variety of small tactical weapons, they conducted many tests of their capabilities. The United States announced most, but not all, of its tests; since the Soviet Union rarely proclaimed its explosions, the United States adopted the role of its adversary's press secretary by announcing some of the Soviet tests it had detected. The reason for this delicacy in publicity was a desire not to allow the other side full knowledge of one's ability to detect very small detonations. Intelligence has always been linked to security, and no more so than in connection with nuclear weapons.

During the 1950s, the superpowers conducted 222 announced nuclear explosions (both fission and fusion), only 18 of them underground or underwater, and more and more frequently newspapers carried stories that scientists in one community or another detected unusual levels of radioactivity in surface water, milk, or children's teeth. In the early days, when smaller weapons were tested, radioactive fallout was known to be carried by the winds, but this was believed to be only a local health problem. [27;49–56; 102:416]

After the March 1, 1954, U.S. test at Bikini of a 15-megaton thermonuclear device, however, global concern was expressed. Unpredicted winds deposited lethal doses of radioactive ash on Rongelap Atoll, 100 miles away, causing emergency evacuation of the natives. Ash fell also on the open sea, where a Japanese fishing boat, the *Lucky Dragon*, trolled outside the posted hazardous area. Crew members became sick (one died), and the catch was found to be radioactive upon the boat's return to port. As more and more large bombs were detonated, the extensive reach of fallout, carried into the stratosphere and there circulated worldwide, became apparent. [70]

The danger of radioactive fallout emerged as a political issue, with Democratic presidential candidate Adlai Stevenson in 1956 condemning testing in the atmosphere. Ironically, Eisenhower had to defend his administration's testing, though he personally wanted to end the practice. Around the same time, chemist Linus Pauling became a prominent advocate of a test ban, for which he was both branded a Communist sympathizer by the Senate Internal Security Subcommittee and awarded the Nobel Peace Prize. [83:257–92]

Citizen activist groups organized to protest radioactive "contamination without representation," among them the National Committee for a Sane Nuclear Policy (SANE) in the United States and the Campaign for Nuclear Disarmament in Great Britain. Their fears of somatic and genetic damage were exacerbated by news of a new weapon design in which the bomb casing was made of "waste" U-238, which does fission under fast-neutron bombardment, though not as well as U-235. These were called FFF bombs, for the fission trigger, the fusion burn, and the casing fission. Use of the sheath in this fashion roughly doubled the explosive yield but made the bombs very "dirty" (in radioactivity) weapons. Efforts by the "peaceniks," as well as changing government policy, led first to a testing moratorium in 1958, and then to the 1963 Limited Test Ban Treaty. [66; 103]

The 1958 moratorium was no more than a gentleman's agreement for a year, extended indefinitely under the terms: "If you don't test, I won't test." Before it began, both nations rushed to complete numerous experiments in the pipeline. The Soviets fired off a range of weapons at their two proving grounds: Semipalatinsk in Kazakhstan and Novaya Zemlya in the Arctic. The United States showed its sophistication by detonating weapons smaller than one kiloton and larger than a megaton, and doing it aboveground and underground in Nevada, as well as 300 miles in space. [55:449–88]

The moratorium was strained by the Soviet shooting down of an American U-2 spy plane in the spring of 1960, and by John F. Kennedy's "missile gap" charges in the presidential campaign later that year. The Central Intelligence Agency's U-2, a glider-like jet aircraft that normally flew above the range of surface-to-air missiles, photographed large swaths of the Soviet Union to detect nuclear missile-launch sites. The plane's destruction maintained the old definition of national airspace: whatever a nation can defend. However, by the time (in the 1980s) that the super-powers developed the capability of shooting down overflying satellites, space was regarded as an international realm.

Kennedy's charge that the Eisenhower administration had allowed the Soviet Union to surpass the United States in the number of nuclear warheads deliverable by missile was good election-year rhetoric but untrue. Both nations had a number of intermediate-range ballistic missiles, while the Soviets had at most half as many ICBMs as the ten Atlas missiles in the U.S. arsenal. American intermediate-range missiles placed in NATO countries could reach the Soviet homeland. Nuclear-powered submarines were also developed in the 1950s, but their major role as platforms for firing SLBMs was a product of the next decade. Through the 1950s, the source of America's overwhelming power was its large fleet of long-range bombers. [54]

These planes of the Strategic Air Command (SAC) were regarded as a deterrent and retaliatory force. Although the United States always refused to commit itself to a no-first-use policy, it seemed inconceivable to most people that America would begin a nuclear war (this despite the attacks on Hiroshima and Nagasaki, and numerous later calls to employ nuclear weapons). [36] The Soviets, of course, looked upon SAC as a potentially aggressive force.

In August 1957, the Soviet Union announced the first successful launch of an ICBM, meaning that it had the ability to build increasingly powerful rockets. In October of that year they showed another use of such rockets by lofting Sputnik, the first artificial earth-orbiting satellite. Americans who felt that the Soviets were technologically backward, and who therefore claimed that their A-bomb and even H-bomb were made possible only by stealing U.S. secrets, were silenced by this major Soviet victory in the "space race."

Sputnik did more than bruise American egos. It meant that Soviet ICBMs could destroy SAC aircraft sitting on the ground. A bomber attack across intercontinental distances would take hours, giving enough radar-warning time for SAC to become airborne. Missiles traveling the same distance would take but half an hour, with only perhaps 15 minutes left after they were detected by radar. SAC thus began an expensive program of keeping a fraction of its fleet in the air at all times, its endurance assisted by aerial refueling. The goal was to maintain a viable nuclear force, even if the home bases were destroyed. The retaliatory ability was therefore credible, presumably ensuring deterrence. (One might note that both sides engaged in a degree of bluffing, for maintenance and other problems often placed aircraft and missiles temporarily out of service.)

SAC bombers would fly to a "hold" position. If sent a properly coded command, they would then fly to their Soviet targets; if no signal or a false signal was received, they would return to base. This "fail-safe" system was controversial, inspiring critics to devise scenarios in which unintended war was ignited by mechanical or electronic errors (the novel *Fail-Safe*), or by a madman (the film *Dr. Strangelove*). Scientists, in their new role of nuclear war analysts, were often the experts who decided what was safe and what was not. As such, they figured prominently as heroes and villains in the novels and films of the period.

With the growth of satellite warning systems and missile capabilities on both sides in the mid-1960s, the airborne alert diminished in importance. Accidents further decreased support for the system. In one, a B-52 and its tanker collided over the Spanish coast near the town of Palomares. In another, a B-52 crashed in Greenland. The several locking mechanisms prevented a nuclear explosion (though some locks failed), but the navy and air force went to great trouble to locate and recover the bombs. The airborne alert soon ended, replaced by having planes and crews ready to take off in a few minutes.

In September 1961, the Soviet Union broke the moratorium on testing bombs, and the United States quickly followed suit. Premier Khrushchev boasted of a 100-megaton weapon, far larger than anything yet produced, and the next month his weaponeers detonated one of 58-megatons. This could well have been the 100-megaton FFF monster, mentioned above, with its fission casing removed to minimize dusting their own nation with massive amounts of radioactivity. This design, in which the casing ordinarily fissions from the neutrons released in the fusion reaction, roughly doubles the yield and was incorporated in many of the warheads of both superpowers.

Fear of nuclear warfare led to civil defense activities. At its minimum, this was simply "duck and cover" exercises for school children: hide under the school desk if a bright flash of light was seen. Evacuation of cities was also contemplated, and the interstate highway system saw its origins in the Eisenhower administration as a means of allowing people in automobiles a faster exit from urban areas. The highway lobby kept the system growing long after civil defense authorities largely abandoned evacuation as unworkable, because a few auto breakdowns could clog the arteries. Shelters were another defensive concept. These were to protect people against fallout, for blast shelters were too expensive (unless subway tunnels were used). Some homeowners constructed shelters for their families, amid discussions about the morality of denying one's unsheltered neighbor access to one's limited space and provisions. The federal government identified deep basements of commercial and public buildings as shelters and stocked some with blankets, cots, crackers, medical supplies, and other items for use during the weeks those sheltered would remain inside, waiting for radioactivity outside to drop to tolerable (whatever that was defined to be) levels. [116]

In October 1962, Premier Khrushchev installed a number of intermediate-range ballistic missiles in Cuba. Their reach encompassed almost all of the 48 contiguous states. President Kennedy demanded that the missiles be removed. The Soviets offered a trade: they would withdrew their weapons if the United States likewise pulled its intermediate-range rockets from Turkey. Although the American weapons were considered obsolescent and there were plans ultimately to remove them, Kennedy refused to bargain overtly. It was for him a matter of both principle and power politics. The United States imposed a naval blockade of the Caribbean island and mobilized land and air forces. Nuclear conflict seemed imminent. Khrushchev then "blinked," agreeing to pull back in response to a United Nations resolution, not Kennedy's demand. It was a face-saving solution to avoid a war that no one wanted and that would have stretched Soviet supply lines to their limit. The Cuban Missile Crisis, as the incident was named, showed the American people the limits of civil defense. They could and did engage in panic buying of food supplies for a feared period of containment in their homes, but there was no time to order construction of new personal fallout shelters. [42]

This crisis taught other lessons, too. The Soviets smarted at the worldwide perception of their inferior strategic strength. They regarded themselves as a superpower and expected to be treated as an equal by the United States. American secretaries of defense and other officials, however, did little to soothe this competitive attitude; instead, they customarily reported on the superiority of U.S. weapons. This may have been a necessary tranquilizer for domestic political purposes, but it had an undesirable hypertensive effect in the Kremlin. The Soviets decided that they would not be humiliated in the future; their arsenal would match that of the West.

The Cuban Missile Crisis, thus, was the watershed of Soviet determination to achieve nuclear parity with the United States. At this time, the United States had about 200 ICBMs, of the Atlas and Titan liquid-fueled types, and the newer solid-fueled Minuteman I. The Soviet Union owned about 50, all liquid-fueled. In addition, the United States boasted of about 100 ballistic missiles on nuclear-propelled submarines, while the Soviets had none. The United States also had 1,700 long-range bombers, all jets, to the Kremlin's 200, which included many turboprop Bear aircraft.

All types of missiles grew in the following decades, while the American bomber fleet shrank. The Soviets did eventually lead in several weapons categories, particularly ICBMs. But the quantities on both sides often did not seem rational. One assumes that threat analyses, strategic goals, and other foreign policy inputs played a part in determining the force structures of both countries, yet the public perception was that "bean counting" was the primary determinant. Even when nuclear warheads were numbered in the tens of thousands, a condition widely regarded as excessive (or "overkill"), far too many politicians and commentators still argued that one side was "ahead" or "behind" in the arms race based on the mere quantity of weapons.

The presentation of complex problems in simple-minded terms is not limited to the arms race, of course. But it certainly made it easy for people to state their positions, more emotionally than pensively, as on bumper stickers reading "Better dead than red" or "One nuclear bomb can ruin your day." President Eisenhower understood better than most the complex, interlocking facets of the arms race and the ways in which it penetrated and permeated American society. In earlier times, military hardware was built in government-owned navy yards and arsenals. While civilian industry expanded production during wartime, it reverted to its customary peacetime size after the hostilities. Not so after World War II. Defense industries flourished as Cold War pressures led to a continuous flow of orders for one weapon system after another. This practice was encouraged by veterans' organizations, politicians who wanted jobs and an economic boost for their constituents, corporations that judiciously spread their subcontracting into the maximum number of congressional districts, and universities that sought to grow in wealth and prestige for the research accomplished on campus. As mentioned above, the situation was aptly called a "permanent war economy." In 1961, in his farewell radio and television address to the nation, Eisenhower voiced his concern about this distortion of American society:

> This conjunction of an immense military establishment and a large arms industry is new in the American experience. The total influence—economic, political, even spiritual—is felt in every city, every State house, every office of the Federal government. We recognize the imperative need for this development. Yet we must not fail to comprehend its grave implications. Our toil, resources and livelihood are all involved; so is the very structure of our society.
>
> In the councils of government, we must guard against the acquisition of unwarranted influence, whether sought or unsought, by the military-industrial complex. The potential for the disastrous rise of misplaced power exists and will persist. [35:98–101]

"Military-industrial complex" became a familiar term in America. The Soviets had their counterpart, but in that closed society it received far less public attention. While widely noted, Eisenhower's warning went largely unheeded, until the dissolution of the Soviet Union three decades later. And even then, the reduction of the military-industrial complex entailed painful political and economic dislocations in both superpowers.

Strategy

Most weapons have bearing upon the outcome of battles, and thus are called tactical weapons. Tactics involve deciding when and how to employ the devices. It does not make sense, for example, to release poison gas from ground-based cannisters if the enemy is upwind. From their invention, nuclear weapons were regarded as so powerful that they would turn the tide not of battles but of the war. Even after small, tactical nuclear rockets, shells, land mines, and other arms entered the arsenal, it was hard to think of them solely as battlefield weapons. To avoid defeat, a commander would likely release his "nukes." His opponent would respond in kind. Back and forth the nuclear weapons would fly, sooner or later including those of greater yield and range. Most scenarios described the superpowers inevitably drawn into a conflict involving long-range weapons. If nuclear war erupted—even if very small at first—it was widely perceived that it would escalate into a full-scale confrontation. Thus, nuclear exchanges were usually seen as strategic in nature, and to be avoided. From the perspective of many people, nuclear weapons were the first weapons made solely to deter and not to be used. [28:30–31]

During World War II the Royal Air Force and the U.S. Army Air Corps attempted to bomb German ball-bearing factories, oil refineries, transportation hubs, and other specific facilities that would destroy the enemy's ability to wage war. In something of a departure from the strategy of many previous wars, where the object was to incapacitate the enemy's troops, a nation's economy was targeted. But bombing inaccuracy and German defenses forced first the British and then the Americans to alter their plans from pinpoint aiming to an area-bombing policy. Making a virtue of this necessity, both Britain and the United States proclaimed the value of their massive air raids of high explosives and incendiary bombs. By shattering enemy workers' homes, they argued, they destroyed their morale. There was debate about the morality of such a doctrine, and postwar surveys questioned whether these means achieved their intended ends, but all agreed that the term "strategic bombing" had come to mean the destruction of cities. [91:131–56]

After the war, when the Soviet Union replaced the Axis powers as the focus of enmity, strategic bombing was regarded as the counterbalance to the huge Soviet armies that threatened Western Europe. This was but a modification of World War II policy, now with Soviet industry, transportation centers, and armed forces that had the potential of striking America (i.e., long-range bombers) as the primary targets. It differed from World War II in that conflict was expected to be over quickly (a "spasm war"), the damage would be more intense, and the cost would be lower (since it would not be necessary to mobilize large armies). One hundred nuclear weapons of Hiroshima size were considered to be an adequate deterrent; the Soviets would be unwilling to accept such losses. In general, cities were not specific objectives, but they would be strongly affected since two of the three types of targets were associated with urban areas. Civilians thus would be counted as "collateral damage." [91:131–56]

Initially, the Soviet air force was not viewed as a significant threat to the American air force. But by the early 1950s, with nuclear weapons in their bomb bays, even a small fleet of aircraft could place American bases in the continental United States and overseas in jeopardy. With conventional weapons, the attacker could not win if he lost 10 percent of his aircraft in each raid. But with nuclear weapons, the defender could not win even if he shot down 90 percent of the attacking aircraft. This startling vulnerability is one of the major changes that sets nuclear weapons apart from conventional explosives. It explains the continuing search for air-, missile-, and civil defense systems, and even more the research and development of new and varied weapons (such as mobile, land-based missiles) that will survive a surprise attack.

When thermonuclear weapons were developed in the 1950s, both opponents and proponents assumed that they would be used primarily against large cities. The inaccuracy of the first generation of long-range missiles made that a reasonable conjecture: only cities were large enough to assure a "hit." In fact, thermonuclear weapons allowed the development of lightweight, low-yield explosives, fashioned into tactical weapons and multiple warheads for a single missile. But the early missiles did carry high-yield warheads, and so too did SAC bombers. Henry Rowen, former president of the RAND Corporation, reported that between 1954 and 1960 the number of delivery vehicles in the U.S. strategic arsenal remained fairly constant around 2,000, but the megatonnage carried increased over 20 times, peaking in 1960. Subsequently, the U.S. megatonnage dropped, as weapons were packaged in smaller sizes for missile nosecones. [91]

But the introduction of long-range missiles reawakened fears of vulnerability to surprise attack. To make their deterrents credible, both superpowers in the 1960s built a "triad" of strategic forces, placing nuclear weapons in bombers, atop land-based ballistic missiles, and atop submarine-launched ballistic missiles. The attacker would be unable to locate and destroy all these forces simultaneously, leaving himself open to inevitable retaliation. Once each side recognized this stalemate of Mutually Assured Destruction (MAD), the strategic balance became dynamically stable.

But strategy is more than the evolution of weapons; it involves geopolitical concepts of their exploitation. If, as stated above, nuclear weapons were made not for use but for deterrence, and if deterrence is to be credible, a nation must posture as if it intends to employ the weapon. Even uncertainty is acceptable as a position, for it leaves the enemy guessing. Thus, with the advent of nuclear weapons, planners sought to incorporate them into foreign and military policies. A great many concepts were suggested, including such relatively inexpensive ones as minimal deterrence (building only enough weapons to deter an attack), arms control, and disarmament. However, the three main strategies actually pursued were called containment, massive retaliation, and flexible response, and they were expensive. [109]

The wartime alliance between the United States and the Soviet Union,

nurtured by Roosevelt at Teheran and Yalta, was fast becoming unraveled by the time of Potsdam. Soviet refusal to leave Iran in 1946,
and the threat of Communist takeovers in Greece and Turkey in
1947, ended any remaining hopes for common goals. The foreign
policy adopted by the United States to deal with aggressive communism
was called containment. The many military bases that were built to
encircle the Soviet Union were to be used as springboards for U.S.
(or NATO) forces to push the "red menace" back within the confines
of its own empire. Nuclear weapons clearly were among the tools to
achieve this.

Containment was reasonably successful in thwarting Moscow's designs
on Western and southeastern Europe, but it could not prevent the
Communist victory in China in 1949, nor of course could it stop Soviet
scientists from creating nuclear weapons in the same year. Truman
had upset Thomas Dewey in the 1948 presidential campaign, but the
country's continuing political division was underscored by unremitting
criticism of his foreign policy. The nation united behind the president
at the outset of the Korean War in June 1950, but once that conflict
became a gory impasse, the Democratic administration was again
perceived as inept in dealing with the Communists.

Containment thus became a major issue in the 1952 presidential
campaign, which pitted war hero Dwight Eisenhower against former
Illinois governor Adlai Stevenson. To Republicans, the millions of people
in China and Eastern Europe who were swept under Communist rule
were clear evidence of containment's failure. In a foreign policy plank
authored by John Foster Dulles, who became Eisenhower's secretary
of state, the Republicans advocated a strategy of liberation. Short of
encouraging rebellions that were doomed to be crushed, active measures
would be taken to encourage captive nations to break free from
Communist domination. Stevenson argued that this policy would lead
to ill-fated revolutions, as indeed did occur in Hungary in 1956 and
Czechoslovakia in 1968. [89]

Liberation might be a suitable concept for Soviet satellites, but it
had little meaning for the Soviet Union itself. Eisenhower felt that
containment was militarily ineffective and economically disastrous;
it spread costly forces too thinly around the Soviet perimeter. For
greater efficiency and a reduced defense budget, he proposed to
respond with massive force not necessarily at the point of Communist
aggression but at a place of his own choice. This massive retaliation
plan was widely attributed to Dulles, though it had been proposed
in 1948 in the so-called Finletter Report to Truman. Since massive
retaliation coincided with the development of the hydrogen bomb, it

was widely perceived to call for high-yield nuclear attacks against Soviet cities. [89]

The Soviet Sputnik and ICBM test launches in 1957 brought home the vulnerability of the U.S. mainland to attack. Given this condition, massive retaliation was no longer a desirable strategy, for nuclear weapons were less and less a unilateral resource of the West. America's technological superiority was diluted, and the balance of terror was more truly a balance. No longer was it an easy certainty that the United States would be able to retaliate against Soviet aggression; now the United States would have to work hard to retain that capacity. To keep the credibility of the deterrent, SAC instituted its airborne alert, and the Pentagon sought to build an ICBM force. The goal was a second-strike capability, meaning the ability to inflict unacceptable damage upon the enemy even after absorbing his attack. Neither side now had a first-strike capability, the ability to attack without suffering significant damage in return. But what was "unacceptable"? Depending upon one's definition, unacceptability occurred as soon as a single SAC bomber was airborne around the clock, or not until ICBMs began to accumulate in the early 1960s. Was destruction of a single city unacceptable, or need the disaster be multiplied?

It would be wrong, however, to focus too much upon cities. For most of the Eisenhower administration, the preferred targets were industrial plants, offensive military forces, and government headquarters, be they near or far from population centers. Soviet urban areas became even less desirable as targets once American cities became more vulnerable to missile attack. For Eisenhower the key theme was that almost any prolonged Soviet intrusion called for a massive nuclear response. [91]

The uneasiness induced by Soviet long-range missiles late in the Eisenhower administration led to a change in strategy once Kennedy was inaugurated in 1961. Kennedy and his secretary of defense, Robert McNamara, wished to avoid the stark choices of suicide or surrender if nuclear deterrence miscarried. Thus, the option of conventional forces loomed much larger in their thinking. Conventional tools would also enhance deterrence, since their employment was far more logical than nuclear weapons in minor confrontations. For these reasons, flexible response replaced massive retaliation as the nation's principal strategy, and in one variant or another it remained dominant to the end of the Cold War. In the early 1960s, this involved the buildup of conventional forces at the same time the United States constructed Polaris submarines (each carrying 16 missiles), Minuteman ICBMs, and a variety of tactical nuclear weapons for NATO. Nonetheless, the strategic budget declined, as an assortment of bombers, short-range missiles, air defense, and other projects were canceled. [21; 91]

Role of the Scientist

The widespread fear that communism would subvert democratic institutions became paranoia once the Soviet Union joined the nuclear club. The Soviets now had the means to inflict their will. It was bad enough for teachers, movie-script writers, and other opinion makers to be suspected of harboring Communist beliefs, but it was a far greater danger if American scientists, those who gave the nation the ultimate weapon, were also tools of the Kremlin. Threatened Communist success in a few democratic European elections highlighted the power of Communist ideology, leading to mandatory loyalty oaths in some American professions and legislation that restricted the flow of individuals and ideas.

Investigations by the House Committee on Un-American Activities (HUAC), the Senate Internal Security Subcommittee, and numerous committees in state legislatures turned up many people who were charged with being "pink," if not "red." Although membership in the American Communist party in the 1930s was not only legal but somewhat fashionable, that did not prevent the destruction of lives and careers in the late 1940s and 1950s, especially if the accused refused to denounce others. In many cases, unwillingness to incriminate oneself, protected by the Fifth Amendment to the U.S. Constitution, was grounds for dismissal from one's employment. To protect itself from the ideas of its adversary, the United States adopted its foe's techniques. [95]

From about 1947 to the early 1960s—much longer than the years dominated by Senator Joseph McCarthy—the United States engaged in repression of its own citizens. Foreigners fared as badly. If we focus upon scientists, we find that many were prevented from traveling abroad by denying them their passports; foreign scientists were barred from attending professional meetings, lecturing, or accepting jobs in the United States by refusal to issue them visas. Scientists whose past or present political beliefs were liberal, socialist, or Communist became the targets of inquisitors, hate-mongers, and the Cold War fearful, whose comforting goal was uniformity of political thought.

In addition, there were restrictions on the exchange of information, spy revelations, and security clearance hearings. Not past actions but future possibilities were at the core of the problem. The customary presumption of innocence was stood on its head; doubts as to loyalty were resolved in favor of the state. Indeed, in the pursuit of "absolute security," it was inevitable that doubts about an individual would always be resolved in favor of the state. Albert Einstein, in acknowledging the award of a membership card in a plumbers' union in 1954, wrote: "I would rather choose to be a plumber or a peddler in the hope to find that modest degree of independence still available under present circumstances." [33:292]

From the recognition in World War II that science was vital to national security, scientists had tangible power. They could give away "secrets," particularly those of nuclear weapons. Although no American scientist was ever convicted of this sort of treason, or even seriously accused, the public was uneasy. They were happy to have the benefits of science and technology, but fearful that the "longhairs" might release a "Frankenstein." The destruction of Hiroshima and Nagasaki, the diabolical experiments of the Nazi doctors, and such novels as *Brave New World* and *1984* produced anxiety about physical harm and conjured up in their minds suspicions about the technocratic organization of society. In a few short years, scientists went from the wizards who made the bomb to security risks.

Theoretical physicist Edward U. Condon was the best-known government employee to be accused of disloyalty. While he was director of the National Bureau of Standards in 1948, the House Un-American Activities Committee called him "one of the weakest links in our atomic security." [108:238] No evidence was presented, the committee being content to rely upon "appearances," such as his association with alleged Communists and with foreigners, and his links with an alleged Communist-front group, the American-Soviet Science Society. With repetitions of the charge, lots of innuendo, leaks to the press, and anonymous letters to the FBI, it appeared that a case was made against him, yet he was repeatedly denied permission to appear before the committee. Observers of the scene thought that HUAC might be trying to attack science; they were certain that HUAC wished to embarrass the Truman administration. Eventually Condon felt that his usefulness at the Bureau of Standards was eroded, and he resigned. He also had to resign his next post, as director of research for the Corning Glass Company, since he was denied security clearance for military projects and felt that the new Eisenhower administration would not give him adequate backing. As a sign of support for Condon, the American Association for the Advancement of Science, the nation's largest scientific organization, elected him its president. [108]

Biochemists Linus Pauling and Salvador Luria were denied their passports in 1952, and thus prevented from attending a meeting on the structure of proteins held at the Royal Society of London. Herman

J. Muller was refused permission by the AEC to attend an International Conference on the Peaceful Uses of Atomic Energy, held in Geneva in 1955; his paper, on human responses to radiation, mentioned Hiroshima, an indelicate word. Foreign scientists who were turned away from American shores included E. B. Chain, Alfred Kastler, George Hevesy, Jacques Monod, and P. M. S. Blackett. Only Nobel laureates are mentioned in this paragraph, but the list of scientists affected by these abhorrent government policies is much longer.

The most notorious case of this era was that of J. Robert Oppenheimer. Told at Christmastime 1953 that the AEC would not renew his security clearance, Oppenheimer fought to clear his name—and lost. His hearing before an AEC personnel security board, in April and May 1954, took on the trappings of a criminal trial. The two main charges were that he kept in contact with family members and friends who had been and perhaps still were Communists (a fact known to General Groves when he selected Oppenheimer to head Los Alamos in 1942), and that at one point he had been opposed to construction of thermonuclear weapons. The most damaging testimony against him was given by Edward Teller, who said that he questioned Oppenheimer's judgment and would prefer that the nation's security was in others' hands (for this statement Teller was ostracized by many in the scientific community for the next decade). While concluding that there was no evidence of disloyalty, but instead an enormous amount of valuable service to his country, the board nonetheless felt that Oppenheimer was a security risk and recommended that his clearance be terminated. The board emphasized his opposition to the H-bomb.

The case then went to the five commissioners of the AEC, where Oppenheimer again lost. Sensitive to the appearance that Oppenheimer was being "convicted" merely for holding an opinion that turned out to be the opposite of national policy, the AEC emphasized not the H-bomb matter but Oppenheimer's "defects of character" (he admitted lying to a wartime security officer whom he wanted to warn about possible Communist penetration of Los Alamos, while at the same time protecting his source, a close friend—another matter passed by Groves). [77]

At the root of the episode was a change in the standards for government employment in sensitive areas. Originally, loyalty and the ability to keep secrets were sufficient. But between 1947 and 1953, the government expanded its criteria and included the concept of security risk. This introduced the measures of character and associations, and the basis for exclusion from government service changed from "reasonable belief of disloyalty" to "reasonable doubt of loyalty." A small change of words, but a great difference in meaning.

There is an epilogue to this story, rather like a Soviet history of purge and rehabilitation. After his humiliation, Oppenheimer returned to his job as director of the Institute for Advanced Study at Princeton, but he seemed a broken man. In 1963, under a different administration and with a lessening of Cold War tensions, Lyndon Johnson presented him with the AEC's highest honor, the $50,000 Fermi Award. Of passing interest to the public, the Oppenheimer "trial" was riveting to the scientific community. Many wondered if scientists would consult for the government in the future if they were treated so shabbily for being on the losing side of a policy disagreement. Fortunately for the government, there was no boycott.

In contrast, far more public attention was given to stories of atomic espionage. And here there were real cases of scientists who also served as spies for the Soviet Union. When Canadian authorities hauled in a soviet spy ring early in 1946, one scientist was in their net: Englishman Allen Nunn May, a physicist who had worked on the development of reactors in Canada. On four occasions in 1944, he visited the Chicago Metallurgical Laboratory and learned something of the plutonium-producing reactors being built at Hanford. His knowledge of bomb developments, however, was marginal. When arrested, he argued that the Soviet Union was an ally in World War II and deserved to receive all possible help. Indeed, May felt that he had made a contribution to the safety of the world. He claimed, further, that his actions were in the long tradition of free exchange of information among scientists. May, like others, became a traitor for ideological, not mercenary, reasons. His sentence of ten years' imprisonment was considered harsh in Britain, where lesser sentences has been given some who had aided the wartime *enemy*, while Americans, wishing only the death penalty for spies, felt it was too lenient. [62].

A far more significant case was that of German-born British theoretical physicist Klaus Fuchs. As a youth, he joined the only organized effort against the Nazis, that of the Communist party, and eventually was forced to flee to England. There he earned a doctorate and was drawn into war work; the British radar project was too sensitive to employ someone who was technically an enemy alien, but their preliminary atomic research was not. Soon after he began investigating the gaseous diffusion process for separating isotopes of U-235, Fuchs contacted Soviet agents and periodically furnished them with documents and reports. When this work was merged with that of the United States, Fuchs was part of the British mission that crossed the Atlantic. Indeed, Fuchs made significant contributions to the success of diffusion at the same time that he described its theory and the Oak Ridge plant, with its porous barrier, stages, and production rate, to the Soviet Union.

When that job was done, he moved to Los Alamos in the summer of 1944, joining the vital theoretical division, under Hans Bethe. There he contributed admirably to the success of the implosion process. He also learned much about the problems in the free-ranging discussions allowed within the laboratory. After his return to England, Fuchs became leader of the theoretical physics division at Harwell, Britain's nuclear establishment. At his trial, he argued that he broke his oath of allegiance to Britain because he had a higher loyalty to humanity, whose safety was enhanced by ending the American nuclear monopoly. Upon conviction in early 1950, he was sentenced to 14 years in prison. Widespread belief that Fuchs had transmitted to Moscow the "secret of the H-bomb" (when in fact no one had yet figured out how to construct it) gave President Truman added justification to embark on the quest for thermonuclear weapons. [62; 117]

May and Fuchs, both of whom confessed, were tried in England. America soon had its own atomic "crime of the century" trial, but it involved no scientists. The defendants were Julius and Ethel Rosenberg, accused not of espionage, because there was no tangible evidence, but of conspiracy to commit espionage, a charge requiring lesser standards to convict. Ethel's brother, David Greenglass, while an army sergeant, had worked as a machinist at Los Alamos, fashioning the high-explosive lenses for the implosion bomb. He, and apparently Klaus Fuchs, provided information to courier Harry Gold, who carried it back to the Rosenbergs in New York, and they in turn conveyed it to the Soviet vice-consul. The Rosenbergs, it was claimed, managed the spy network.

Maintaining their innocence throughout, they were convicted and condemned to death in the electric chair, a sentence carried out in 1953, despite worldwide appeals for clemency. Whether they were guilty remains a controversial question; whether they received a fair trial is not in doubt: they did not. The judge was in constant contact with the prosecution, at best a violation of legal ethics, at worst, collusion. Witnesses for the prosecution, including Greenglass and Gold, had reason to cooperate with the Justice Department, for they were faced with prosecution themselves. Recorded conversations between Gold and his attorneys before the trial suggest that he had no recollection of events he described later in great detail, after long discussions with the prosecution. The FBI inexplicably allowed the original of the document that allegedly proved a conspiracy to be destroyed. A sketch of the bomb was misrepresented as Greenglass's original when it was a replica he drew after his arrest. These and other instances of questionable behavior tainted the trial so badly that it is widely regarded not as a shining example of American jurisprudence but as a symbol of McCarthy-era hysteria. The government seemed to need a conviction to reassure the public that atomic secrets were being kept safely. The Rosenbergs may in fact have been guilty, but their trial did not prove it. [94]

Most scientists, of course, did nothing that would lead them into such sensational courtroom dramas. Yet many of them found new public roles to play that took them from their laboratories for longer or shorter periods and on occasion did thrust them into the limelight. After several years of effort, Congress in 1950 passed legislation creating the National Science Foundation (NSF), which distributed federal funds for nonmilitary research to academic scientists. Scientists served as administrators and as members of its governing National Science Board (which meant that scientists were now on top, as well as on tap). The Department of Defense (DOD), through such agencies as the Office of Naval Research, and the Atomic Energy Commission continued to fund research, with budgets far larger that that of NSF. DOD created a Defense Science Board, and the individual services had chief scientists and appropriate staffs. At one point a physicist (Harold Brown, former director of the Livermore Laboratory) was secretary of defense, while a chemist (Glenn Seaborg, discoverer of plutonium) was chairman of the AEC. [37]

The prestige scientists won in World War II allowed them the latitude to comment on subjects foreign to their technical training. The Federation of American Scientists, as described above, campaigned in 1945 and 1946 for its ideal of legislation. In following years, other organizations of scientists were created, including the Union of Concerned Scientists and Physicians for Social Responsibility, while the Pugwash Movement became an international forum where problems could be discussed informally (instead of in negotiations between governments). Such groups usually played the role of loyal opposition, for in general they sought to move the American government in the direction of arms control and an amelioration of the arms race. These goals were specifically those of the Council for a Livable World, a political action committee founded in 1962 by Leo Szilard, which sought to influence foreign policy by supporting the election of liberals to the U.S. Senate.

Not all scientists, of course, were "doves," but the "hawks" had less reason to organize themselves. Scientists of both feather markings, however, served on prestigious advisory boards to the government, such as the AEC's General Advisory Committee and the President's Science

Advisory Committee. The latter was created in response to the anxiety generated by Sputnik; a full-time position of president's science advisor was also established at that time, filled first by James Killian of MIT.

Lower levels of government also found the need for technical expertise, which they acquired from employees and from independent contractors. (It is worth noting that technical reports dealing with public policy issues often carry political implications or "spin"; few documents are value-free.) In government and in a new form of institution—the "think tank"—natural scientists were joined by social scientists, such as psychologists, economists, historians, and political scientists. The RAND Corporation, founded in 1947 to provide policy analyses for the air force, was perhaps the first of a great number of these private companies doing public work. Those engaged in this kind of applied research were sometimes called "defense intellectuals." Few became more famous than Herman Kahn, RAND physicist and author of *On Thermonuclear War* (1960) and *Thinking about the Unthinkable* (1962). Kahn believed that there must be more than just the choices of world war or capitulation, should the Soviet Union assault the West, and described in great and horrifying detail numerous alternative scenarios of superpower engagement. His work meshed well with the inclinations of the military-industrial complex, for a multiplicity of hostile situations required a multiplicity of weapons. [64; 65; 100]

Arms Control

The first arms control proposal of the nuclear age was the Baruch Plan in 1946. Like many subsequent offers by both superpowers, it had acceptable features, unacceptable features, and a healthy dose of posturing for other nations of the world. Its failure at the UN, amid a hardening of Cold War attitudes, led to a decade-long lull in arms control efforts, during which both the United States and the Soviet Union concentrated on building their fission and fusion arsenals. That very process, however, led to the testing moratorium of 1958, when worldwide protest against radioactive fallout was too strong to be ignored. In addition, it was becoming increasingly obvious that enormous amounts of national treasure were being spent on arms, without making populations more secure than before.

This dilemma was obvious to Soviet and American leaders, who occasionally showed signs that they wished to escape from it. They were fettered, however, by domestic political realities, which precluded anything that looked like appeasement. Under these constraints and those of mutual suspicion and distrust, negotiations proceeded intermittently from the mid-1950s, hampered further by such Cold War events as the downing of the U-2 spy plane over Russia, erection of the Berlin Wall, the attempted invasion of Cuba at the Bay of Pigs, and the Cuban Missile Crisis. For most of these years the focus was upon a comprehensive test ban (CTB) treaty, outlawing all tests. Since neither side intended to sign a pact that involved trusting the other, the ability to verify that the treaty provisions were honored was vital. But the science of seismology was not well enough developed to distinguish, from beyond a nation's borders, small underground explosions from earthquakes. Some on-site inspections of "suspicious events" were required. [22]

In a departure from the Soviet view that equated inspection with spying, Premier Khrushchev offered to permit two or three on-site visits a year. Since the United States and Britain had lowered their demand to six inspections, it seemed as if a CTB was near agreement. Yet no CTB accord was signed, each side having already reached its political limit. Indeed, they really were further apart than it appeared, for the Soviets could not agree to the Western concept of an on-site inspection: a helicopter survey of 500 square kilometers and the drilling of many holes for earth samples. That appeared too much like spying. [29; 96]

The Cuban Missile Crisis of October 1962 unnerved both Khrushchev and Kennedy. To both, nuclear war meant national suicide, and we had come close to it. Kennedy was additionally motivated to conclude a test ban, for he feared the proliferation of nuclear weapons to other countries. If a CTB still could not be achieved, why not finesse the problem of underground testing by stepping around it? Thus, a limited test ban treaty began to look more like a prize worth having. It would ban testing in the atmosphere, in space, and under water. This had the advantage of being far more acceptable to the U.S. Senate and the Joint Chiefs of Staff. Thus, the Limited Test Ban Treaty was signed by the foreign ministers in August 1963. To achieve ratification by the Senate the following month, Kennedy had to promise to pursue a vigorous program of underground testing, maintain the expertise and quality of the Los Alamos and Livermore labs, and support an expansion of seismological research to perfect detection of underground tests. [29; 96]

Although France and China declined to sign the Limited Test Ban Treaty and tested their weapons in the atmosphere, the pact did reduce fallout significantly, for the three largest nuclear powers went underground. Equally important, it gave momentum to arms control negotiations, showing that they could be accomplished successfully.

Accords reached later addressed Outer Space (1967), Latin America (1967), Non-Proliferation (1968), and the Seabed (1971). The first Strategic Arms Limitation Talks (1972, abbreviated as SALT I) set the bounds of certain classes of weapons but did not reduce the arsenals. SALT II (1979, but not ratified by the US Senate) set somewhat smaller limits but again involved virtually no reductions. The Intermediate Nuclear Forces Treaty (1987) was the first to go beyond merely capping the number of existing weapons to mandate the destruction of a whole class of weapon.

Finally, the Strategic Arms Reduction Talks (START) and negotiations between the United States and Russia in mid-1992 (which amazingly had no acronym at first, but came to be called START II) achieved significant reductions in strategic weapons. Indeed, the latter set limits (3,000–3,500 warheads) that were about one-third the number each side had in its arsenal. Such a goal, unthinkable but a short time before, was due to the demise of the Soviet Union. Now, as the United States and Russia reconfigure their nuclear forces in light of the new numbers and new political realities, there is rekindled interest in the concept of minimum deterrence. Nuclear weapons will not be completely foresaken—unless a more exotic form of destruction is developed—but minimum deterrence advocates believe that a thousand, or even just a few hundred, warheads should suffice. The Limited Test Ban Treaty, thus, did not end the arms race, but it marked the beginning of a trend that ultimately led to the end of the Cold War.

Epilogue

The primary focus of this book ends in 1963, but the arms race and the Cold War which drove it went on for almost three decades more. Geopolitical flashpoints included the Soviet invasion of Czechoslovakia (1968), the US war in Vietnam (1961–75), and the Soviets' own "Vietnam" in Afghanistan (1979–89). Weaponry continued to make great strides during this period. Both nations developed their triad of strategic forces—ICBMs, SLBMs, and aircraft-dropped bombs—with the United States placing the majority of its warheads on difficult-to-target submarines, and the Soviet Union, traditionally a land power, keeping most of its warheads (including those on mobile missiles) within its borders. Nonmobile, land-based missiles were buried in underground silos, whose walls were hardened with reinforced concrete to withstand the great overpressure of the ever more accurate missiles of the enemy. Indeed, the measure of accuracy of an ICBM is now as small as 300 feet, meaning that the target will likely be within the explosion's crater.

Multiple independently targetable reentry vehicles were built beginning in the late 1960s, and antiballistic missiles (ABMs) were constructed in the early 1970s, though the low confidence in their ability to shoot down an incoming ICBM led to the SALT I treaty, which limited their deployment. The Soviets completed an ABM system to protect Moscow, while the United States inactivated the one complex it erected to protect the ICBMs at the Grand Forks Air Force Base in North Dakota. The Strategic Defense Initiative (SDI, also called "Star Wars"), launched by President Ronald Reagan in 1983, was a later version of an attempt to design an ABM system. Many scientists, frustrated by the political rejection of their technical advice, went public with their opposition to the ABM around 1970 and were credited with a major role in that system's demise. Similar widespread criticism of SDI by scientists—for example, a study conducted by the American Physical Society—reduced public support for the program but did not cause its end.

President Richard Nixon, widely regarded as an arch cold-warrior, nonetheless recognized the economic insanity of trying to stay quantitatively ahead of the Kremlin in all aspects of weaponry. "Sufficiency" replaced "superiority" as the guideline, at least for a while. Although new bombers were developed (the B-1 and the "stealth" B-2, whose construction offered low radar reflection), the air force's B-52 bombers remain the workhorses of the aircraft portion of the triad, with their usefulness extended by the development of cruise missiles. The cruise missile, which can be launched from the B-52 at a safe distance from a defended target, is essentially a pilotless jet airplane, which flies at radar-evading treetop level.

The Soviets, whose style was to build smaller numbers of more varieties of ICBMs than was the U.S. custom, constructed some very large ones, such as the SS-18. Its capability of carrying numerous MIRVed warheads was matched by the U.S. MX missile. The MX, however, was designed to be mobile to make it a difficult target, but every proposed plan to achieve that mobility was defeated because of environmental impact, the unwillingness of populations to have such weapons traverse urban centers on road or rail, or some other essentially nonmilitary consideration. Congress finally slashed the number originally requested, and Reagan ultimately planted those built in old Minuteman silos, which of course are stationary.

MIRVing missiles increased the number of warheads enormously. It also made a MIRVed missile a more appealing target; one incoming warhead could destroy a missile with perhaps ten warheads, while it was still in its silo. In a tense situation, one side might seek to attack first, to minimize the other side's retaliatory capability. To reduce this incentive for a first strike, development of the single-warhead, road-mobile Midgetman ICBM was begun. Its mobility would reduce its vulnerability, and its single warhead would reduce an enemy's interest in it. Midgetman was intended ultimately to replace Minuteman and MX ICBMs. With the end of the Cold War, however, President George Bush canceled both MX and Midgetman programs. Further, MX missiles already deployed will be taken out of service as a consequence of the treaty concluded by Presidents Bush and Yeltsin in 1992. MIRVed missiles—those on land, at any rate—are regarded as destabilizing; they are more threatening and more threatened than missiles having but one warhead. Many of these will be decommissioned as a result of the START II agreement, which will lead to the shrinking of strategic weapon numbers by about two-thirds. The navy's Polaris- and later Poseidon-missile-launching submarines, with 16 tubes each, are being replaced by the 24-tube Trident submarine, which carries MIRVed missiles of the same name. The Trident's notable features are a missile range and accuracy comparable with that of the MX.

While the United States and Russia still hold numbers of nuclear weapons, and Great Britain, France, and China also have significant arsenals, conflict between any of these nations now seems remote. With the end of the Cold War, greater attention is focused upon the problem of proliferation, especially to Third World countries and terrorist groups. To the long-standing fear that these nations or organizations might steal a weapon or build their own bomb factories, we now add the concern that they might purchase nuclear weapons from the stockpiles of the economically desperate former Soviet Union—or hire its redundant bomb designers to fashion their arsenals. Once again, the public has some reason to fear scientists.

At the start of this story, scientists rarely were seen out of their laboratories. A few worked in industry, in such research facilities as those of the Bell Telephone and General Electric companies, and a few were employed by government, at the National Bureau of Standards and in such agencies as the Bureau of Mines, the Geological Survey, and the Department of Agriculture. When scientists made headlines it was usually because of their research results and far less commonly because of their political views. The latter was the case, for example, with the British Social Relations of Science movement of the 1930s, whose members advocated conscious use of science for the benefit of society. Like their counterpart in the United States, the American Association of Scientific Workers, they were tarred with the "Communist" brush, and thus were marginalized. [56; 114]

In contrast, scientists in the postwar period appeared in a number of roles. Besides their normal laboratory activities in universities, they now were "principal investigators" on the contracts and grants received from the government, for only Washington had the funds to support science in the manner desired by both recipients and grantor. The government wisely recognized that the applied-science results it so warmly welcomed were rooted in basic research, so that even result-oriented agencies, such as the military services, happily paid for fundamental inquiries of nature. DOD was not entirely altruistic, it should be added. Its support of basic research enabled it to "stockpile" scientists—to build a pool of expertise that was comfortable in its association with the military—for a time when DOD would need them.

Scientists filled other roles as well. They worked in an increasing array of think tanks and "high-tech" industries. Quickly developing sensitive political antennae, they testified before Congress, lobbied the Senate, House, and administration, consulted for government agencies, and served on numerous boards and committees. They were employed in the several national laboratories and, indeed, worked throughout the federal government. The National Science Foundation, the National Aviation and Space Administration, the Department of Energy, and the Central Intelligence Agency come easily to mind as federal "consumers" of technical talent, but the list is much longer. Not many scientists rose to positions of authority, but some did achieve high levels in government, including that of secretary of defense.

Scientists frequently found it necessary or useful to act as the government's conscience. From its postwar inception to the present time, the monthly *Bulletin of the Atomic Scientists* has been a forum for the advocacy of stronger arms control and environmental protection measures than those generated within the Washington Beltway. Similarly, arms control efforts by Leo Szilard and by Albert Einstein (the Einstein-Russell Manifesto), and opposition to atmospheric testing by Linus Pauling, illustrate the political role increasingly played by scientists. Not only individuals engaged in this effort to mold public opinion; so did organizations such as the Federation of American Scientists, Union of Concerned Scientists, and Physicians for Social Responsibility.

The dark side of scientists' role in the modern world must also be mentioned. Klaus Fuchs and a few others showed that wisdom and ethics are not directly proportional to intelligence. Espionage is, of course, an extreme action, but perfectly legal steps taken by scientist-hawks and by scientist-doves led parts of the public with the opposite political stance to view scientists in general with apprehension. Thus, we find them stereotyped in such artistic endeavors as the popular films mentioned earlier and in Friedrich Dürrenmatt's gripping play *The Physicists*.

The importance of science to modern society is too obvious to be doubted. Among consumer products alone, we can count television, transistor radios, telephone calls over fiber cables, solid-state calculators, computers, microwave ovens, electronic wristwatches, pharmaceuticals, jet aircraft travel, and a host of other "necessities" of life. But this pervasiveness of science is really a product of World War II. The development of radar, the proximity fuse, penicillin, and, most of all, the atomic bomb proved then that science was not merely desirable but vital to national goals. With science, its practitioners rose to prominence. The seeds of their political influence can be traced to such small beginnings as the Einstein letter to Roosevelt and the Franck Report. That influence then blossomed in the many roles that scientists played in the nuclear arms race.

Earlier in the century, scientific (and other professional) societies regarded themselves as scholarly organizations. Their functions were to hold meetings and publish journals, and little else. In the postwar period social responsibility became a stronger and stronger force among American scientists, with each unpopular war fought and with each controversial weapon decision made. One by one, and not without resistance, scientific societies adopted bylaws or policy statements that affirmed their duty, as they saw it, to try to influence national policy. To deal with political issues, some even created sections, such as the American Physical Society's Forum on Physics and Society, or committees, such as the National Academy of Sciences' Committee on Science and Public Policy. Despite occasional criticism that these societies engage in special pleading, such as when they urge the government to increase funding for science, this new role has been of great benefit to the nation. Collectively and individually, scientists have expertise that informs the political process. To ignore their public policy voice would be as misguided as to spurn their technical skills.

Bibliography

[1] Gar Alperovitz. *Atomic Diplomacy: Hiroshima and Potsdam.* New York: Random House, 1965.

[2] Gar Alperovitz. "The Trump Card." *New York Review of Books,* June 15, 1967, 6–12.

[3] Joseph and Stewart Alsop. "Your Flesh *Should* Creep." *Saturday Evening Post,* July 13, 1946, 9.

[4] "America's Atomic Atrocity." *Christian Century,* Aug. 29, 1945, 974–76.

[5] Atomic Energy Commission. *The United States Atomic Energy Commission: What It Is, What It Does.* Oak Ridge, TN: AEC, 1967.

[6] Francis Bacon. *The Great Instauration,* in Edwin A. Burtt, ed., *The English Philosophers from Bacon to Mill.* New York: Random House, 1939.

[7] Francis Bacon. *The New Atlantis,* in Charles W. Eliot, ed., *The Harvard Classics,* vol. 3. New York: P. F. Collier, 1909.

[8] Francis Bacon. *Novum Organum,* sec. 81, in Edwin A. Burtt, ed., *The English Philosophers from Bacon to Mill.* New York: Random House, 1939.

[9] L. Badash. "The Age-of-the-Earth Debate." *Scientific American* 261 (Aug. 1989): 90–96.

[10] L. Badash. "British and American Views of the German Menace in World War I." *Notes and Records of the Royal Society of London* 34 (July 1979): 91–121.

[11] L. Badash. "How the 'Newer Alchemy' Was Received." *Scientific American* 215 (Aug. 1966): 88–95.

[12] L. Badash. *Kapitza, Rutherford, and the Kremlin.* New Haven: Yale University Press, 1985.

[13] L. Badash. "Nuclear Physics in Rutherford's Laboratory before the Discovery of the Neutron." *American Journal of Physics* 51 (Oct. 1983): 884–89.

[14] L. Badash. "The Origins of Big Science: Rutherford at McGill," in Mario Bunge and William Shea, eds., *Rutherford and Physics at the Turn of the Century.* London: Dawson, 1979.

[15] L. Badash. "Radioactivity before the Curies." *American Journal of Physics* 33 (Feb. 1965): 128–35.

[16] L. Badash. *Radioactivity in America: Growth and Decay of a Science.* Baltimore: Johns Hopkins University Press, 1979.

[17] L. Badash, J. O. Hirschfelder, and H. P. Broida, eds. *Reminiscences of Los Alamos, 1943–1945.* Dordrecht: D. Reidel, 1980.

[18] L. Badash, Elizabeth Hodes, and Adolph Tiddens. "Nuclear Fission: Reaction to the Discovery in 1939." *Proceedings of the American Philosophical Society* 130 (June 1986): 196–231.

116   *Bibliography*

[19] Paul R. Baker, ed. *The Atomic Bomb: The Great Decision.* Hinsdale, IL: Dryden Press, 1976.

[20] Hanson W. Baldwin. "The Atom Bomb and Future War." *Life*, Aug. 20, 1945, 17–20.

[21] Desmond Ball. "U.S. Strategic Forces: How Would They Be Used?" *International Security* 7 (Winter 1982–83): 31–60.

[22] Norman Bentwich. *The Rescue and Achievement of Refugee Scholars.* The Hague: Nijhoff, 1953.

[23] Lord Beveridge. *A Defence of Free Learning.* London: Oxford University Press, 1959.

[24] Alan D. Beyerchen. *Scientists under Hitler: Politics and the Physics Community in the Third Reich.* New Haven: Yale University Press, 1977.

[25] Patrick M. S. Blackett. *Fear, War, and the Bomb: Military and Political Consequences of Atomic Energy.* New York: McGraw-Hill, 1948.

[26] Martin Brown, ed. *The Social Responsibility of the Scientist.* New York: Free Press, 1971.

[27] Barry Commoner. *The Closing Circle: Nature, Man, and Technology.* New York: Knopf, 1971.

[28] Department of Defense. *1991, Military Forces in Transition.* Washington, DC: GPO, 1991.

[29] Robert A. Divine. *Blowing on the Wind: The Nuclear Test Ban Debate, 1954–1960.* New York: Oxford University Press, 1978.

[30] Stephen Duggan and Betty Drury. *The Rescue of Science and Learning: The Story of the Emergency Committee in Aid of Displaced Foreign Scholars.* New York: Macmillan, 1948.

[31] A. Hunter Dupree. *Science in the Federal Government: A History of Policies and Activities to 1940.* Cambridge: Harvard University Press, 1957.

[32] John T. Edsall. *Scientific Freedom and Responsibility.* Washington, DC: American Association for the Advancement of Science, 1975.

[33] Albert Einstein. Letter of Nov. 1954, quoted in Gorton Carruth and Eugene Ehrlich, *The Harper Book of American Quotations.* New York: Harper and Row, 1988.

[34] Dwight D. Eisenhower. *Mandate for Change, 1953–1956.* Garden City, NY: Doubleday, 1963.

[35] Dwight D. Eisenhower. *Public Papers of the Presidents of the United States: Dwight D. Eisenhower, 1960–1961.* Washington, DC: GPO, 1961, 1036–40, as rpt. in William H. Chafe and Harvard Sitkoff, eds., *A History of Our Time: Readings of Postwar America.* New York: Oxford University Press, 1987.

[36] Daniel Ellsberg. "Call to Mutiny." *Monthly Review* 33 (Sept. 1981): 1–26.

[37] J. Merton England. *A Patron for Pure Science: The National Science Foundation's Formative Years, 1945–57.* Washington, DC: National Science Foundation, 1982.

[38] Herbert Feis. *The Atomic Bomb and the End of World War II.* Princeton: Princeton University Press, 1966.

[39] Laura Fermi. *Atoms in the Family: My Life with Enrico Fermi.* Chicago: University of Chicago Press, 1954.

[40] Edwin Fogelman, ed. *Hiroshima: The Decision to Use the A-Bomb.* New York: Scribner's, 1964.

[41] "The Fortune Survey." *Fortune*   (Dec. 1945): 303–10.

[42] Raymond L. Garthoff. *Reflections on the Cuban Missile Crisis.* Washington, DC: Brookings Institution, 1989.

[43] Otto Glasser. *Wilhelm Conrad Röntgen and the Early History of the Roentgen Rays.* Springfield, IL: C. C. Thomas, 1934.

[44] Samuel Glasstone. *The Effects of Nuclear Weapons.* Washington, DC: Department of Defense and Atomic Energy Commission, 1962.

[45] I. N. Golovin. *I. V. Kurchatov: A Socialist-Realist Biography of the Soviet Nuclear Scientist.* Bloomington, IN: Selbstverlag Press, 1968.

[46] Margaret Gowing. *Britain and Atomic Energy, 1939–1945.* London: Macmillan, 1964.

[47] Morton Grodzins and Eugene Rabinowitch, eds. *The Atomic Age.* New York: Simon and Schuster, 1965.

[48] Stephane Groueff. *Manhattan Project: The Untold Story of the Making of the Atomic Bomb.* Boston: Little, Brown, 1967.

[49] Otto Hahn. *Otto Hahn: A Scientific Autobiography.* New York: Scribner's, 1966.

[50] John Heilbron and Thomas Kuhn. "The Genesis of the Bohr Atom." *Historical Studies in the Physical Sciences* 1 (1969): 211–90.

[51] Gregg Herken. *The Winning Weapon: The Atomic Bomb in the Cold War, 1945–1950.* New York: Knopf, 1981.

[52] Richard Hewlett and Oscar Anderson. *The New World, 1939–1946: Volume 1 of a History of the United States Atomic Energy Commission.* University Park: Pennsylvania State University Press, 1962.

[53] Richard G. Hewlett and Francis Duncan. *Atomic Shield, 1947–1952: Volume 2 of a History of the United States Atomic Energy Commission.* University Park: Pennsylvania State University Press, 1969.

[54] Richard G. Hewlett and Francis Duncan. *Nuclear Navy, 1946–1962.* Chicago: University of Chicago Press, 1974.

[55] Richard G. Hewlett and Jack M. Holl. *Atoms for Peace and War, 1953–1961: Eisenhower and the Atomic Energy Commission: Volume 3 of a History of the United States Atomic Energy Commission.* Berkeley and Los Angeles: University of California Press, 1989.

[56] Elizabeth Hodes. "Precedents for Social Responsibility among Scientists: The American Association of Scientific Workers and the Federation of American Scientists, 1938–1948." Ph.D. diss., University of California, Santa Barbara, 1982.

[57] Jack M. Holl, DOE chief historian. Letter to the author, early 1982.

[58] David Holloway. "Entering the Nuclear Arms Race: The Soviet Decision to Build the Atomic Bomb, 1939–45." *Social Studies of Science* 11 (1981): 159–97.

[59] "Horror and Shame." *Commonweal,* Aug. 24, 1945, 443–44; and "Atomic Bomb." *Commonweal,* Aug. 31, 1945, 468.

[60] David Irving. *The German Atomic Bomb: The History of Nuclear Research in Nazi Germany.* New York: Simon and Schuster, 1968.

[61] Joint Committee on Atomic Energy. *Current Membership of the Joint Committee on Atomic Energy, Congress of the United States.* 90th Congress, 1st Session. Washington, DC: GPO, Feb. 1968.

[62] Joint Committee on Atomic Energy. *Soviet Atomic Espionage.* 82nd Congress, 1st Session. Washington, DC: GPO, Apr. 1951.

[63] Paul Josephson. *Physics and Politics in Revolutionary Russia.* Berkeley and Los Angeles: University of California Press, 1992.

[64] Herman Kahn. *On Thermonuclear War*. Princeton: Princeton University Press, 1960.

[65] Herman Kahn. *Thinking about the Unthinkable*. New York: Avon Books, 1962.

[66] Milton S. Katz. *Ban the Bomb: A History of SANE, the Committee for a Sane Nuclear Policy, 1957-1985*. New York: Greenwood Press, 1986.

[67] George F. Kennan. "A Plea for Diplomacy." *Harper's* (Apr. 1984): 9-11.

[68] Andrew Kippis . *The Life and Voyages of Captain James Cook*. London, 1788; rpt. London: Newnes, 1904.

[69] David Knight. "Humphrey Davy." *Dictionary of Scientific Biography* 3:603. New York: Scribner's, 1971.

[70] Ralph Lapp. *The Voyage of the Lucky Dragon*. New York: Harper, 1958.

[71] John W. Lewis and Xue Litai. *China Builds the Bomb*. Stanford: Stanford University Press, 1988.

[72] Walter Lippmann. Quoted in Stanley Meisler, "Soviet Union's End Deprives Americans of a Top Attraction." *Los Angeles Times*, Jan. 4, 1992.

[73] M. Stanley Livingston, ed. *The Development of High-Energy Accelerators*. New York: Dover, 1966.

[74] Manhattan Engineer District, US Army. *The Atomic Bombings of Hiroshima and Nagasaki*. Washington, DC: US Army, ca. 1946.

[75] Dexter Masters and Katherine Way, eds. *One World or None: A Report to the Public on the Full Meaning of the Atomic Bomb*. New York: McGraw-Hill, 1946.

[76] Charles L. Mee. *Meeting at Potsdam*. New York: Dell, 1976.

[77] Peter Michelmore. *The Swift Years: The Robert Oppenheimer Story*. New York: Dodd, Mead, 1969.

[78] Byron S. Miller. "A Law Is Passed: The Atomic Energy Act of 1946." *University of Chicago Law Review* 15 (Summer 1948): 799-821.

[79] Louis Morton. "The Decision to Use the Atomic Bomb." *Foreign Affairs* 25 (Jan. 1957): 334-53.

[80] "Opinion: Doubts and Fears." *Time*, Aug. 20, 1945, 36; letters to the editor, Aug. 27, 1945, 2.

[81] J. Robert Oppenheimer. "Physics in the Contemporary World." *Technology Review* 50 (Feb. 1948): 203.

[82] Pacific War Research Society. *Japan's Longest Day*. London: Transworld, 1969.

[83] Linus Pauling. "Science and Peace" (prize for 1962, presented Dec. 1963), in Frederick Haberman, ed., *Nobel Lectures: Peace, 1951-1970*. Amsterdam, London, and New York: Elsevier, 1972.

[84] Rudolf Peierls. *Bird of Passage: Recollections of a Physicist*. Princeton: Princeton University Press, 1985.

[85] Derek J. de Solla Price. *Little Science, Big Science*. New York: Columbia University Press, 1963.

[86] Eugene Rabinowitch, ed., articles by A. P. Alexandrov and Igor Golovin, under the heading "Igor Kurchatov, 1903-1960." *Bulletin of the Atomic Scientists* 23 (Dec. 1967): 8-18.

[87] Robert Reid. *Marie Curie*. New York: Dutton, 1974.

[88] Richard Rhodes. *The Making of the Atomic Bomb*. New York: Simon and Schuster, 1986.

[89] Hugh Ross, ed. *The Cold War: Containment and Its Critics*. Chicago: Rand McNally, 1963.

[90] Theodore Roszak. *The Making of a Counter Culture: Reflections on the Technocratic Society and Its Youthful Opposition.* New York: Doubleday, 1969.

[91] Henry S. Rowen. "The Evolution of Strategic Nuclear Doctrine," in Laurence Martin, ed., *Strategic Thought in the Nuclear Age.* Baltimore: Johns Hopkins University Press, 1979.

[92] S. L. Sanger. *Hanford and the Bomb: An Oral History of World War II.* Seattle: Living History Press, 1989.

[93] Warner R. Schilling. "The H-bomb Decision: How to Decide without Actually Choosing." *Political Science Quarterly* 76 (Mar. 1961): 24–46.

[94] Walter and Miriam Schneir. *Invitation to an Inquest.* Garden City, NY: Doubleday, 1965.

[95] Ellen W. Schrecker. *No Ivory Tower: McCarthyism and the Universities.* New York: Oxford University Press, 1986.

[96] Glenn T. Seaborg. *Kennedy, Khrushchev, and the Test Ban.* Berkeley and Los Angeles: University of California Press, 1981.

[97] Emilio Segrè. *Enrico Fermi: Physicist.* Chicago: University of Chicago Press, 1970.

[98] Martin Sherwin. *A World Destroyed: The Atomic Bomb and the Grand Alliance.* New York: Knopf, 1975.

[99] Alice K. Smith. *A Peril and a Hope: The Scientists' Movement in America, 1945–47.* Chicago: University of Chicago Press, 1965.

[100] Bruce L. R. Smith. *The RAND Corporation: Case Study of a Nonprofit Advisory Corporation.* Cambridge: Harvard University Press, 1966.

[101] Henry D. Smyth. *Atomic Energy for Military Purposes: The Official Report on the Development of the Atomic Bomb Under the Auspices of the United States Government, 1940–1945.* Princeton: Princeton University Press, 1945.

[102] Stockholm International Peace Research Institute. *World Armaments and Disarmament: SIPRI Yearbook 1976.* (Cambridge: MIT Press, 1976.

[103] Richard Taylor and Colin Pritchard. *The Protest Makers: The British Nuclear Disarmament Movement of 1958–1965, Twenty Years On.* Oxford and New York: Pergamon Press, 1980.

[104] Brian Tierney, Donald Kagan, and L. Pearce Williams, eds. *The Cold War: Who Is to Blame?* New York: Random House, 1967.

[105] John Toland. *The Rising Sun: The Decline and Fall of the Japanese Empire, 1936–1945.* New York: Bantam Books, 1971.

[106] United States Strategic Bombing Survey. *The Effects of Atomic Bombs on Hiroshima and Nagasaki.* Washington, DC: GPO, 1946.

[107] Mark Walker. *German National Socialism and the Quest for Nuclear Power, 1939–1949.* Cambridge: Cambridge University Press, 1989.

[108] Jessica Wang. "Science, Security, and the Cold War." *Isis* 83 (June 1992): 238–69.

[109] Arthur Waskow, ed. *The Debate over Thermonuclear Strategy.* Boston: Heath, 1965.

[110] Spencer Weart. *Nuclear Fear: A History of Images.* Cambridge: Harvard University Press, 1988.

[111] Spencer Weart and Gertrud Weiss Szilard, eds. *Leo Szilard: His Version of the Facts.* Cambridge: MIT Press, 1978.

[112] Charles Weiner. "A New Site for the Seminar: The Refugees and American Physics in the Thirties." *Perspectives in American History* 2 (1968): 190–234.

[113]Victor F. Weisskopf. "Nobel Prizes" [Bethe]. *Science* 158 (Nov. 10, 1967): 745–46.

[114] Gary Werskey. *The Visible College.* London: Allen Lane, 1978.

[115] Lynn White, Jr. "The Historical Roots of Our Ecological Crisis." *Science* 155 (Mar. 10, 1967): 1203–7.

[116] Eugene Wigner, ed. *Survival and the Bomb: Methods of Civil Defense.* Bloomington: Indiana University Press, 1969.

[117] Robert C. Williams. *Klaus Fuchs: Atom Spy.* Cambridge: Harvard University Press, 1987.

[118] David Wilson. *Rutherford: Simple Genius.* Cambridge: MIT Press, 1983.

[119] Lawrence Wittner. *Cold War America: From Hiroshima to Watergate.* New York: Praeger, 1974.

[120] John Winthrop. *Relation of a Voyage from Boston to Newfoundland, for the Observation of the Transit of Venus, June 6, 1761.* Boston, 1761.

[121] Herbert York. *The Advisors: Oppenheimer, Teller, and the Superbomb.* San Francisco: W. H. Freeman, 1976.

[122] Herbert York. "The Great Test-Ban Debate." *Scientific American* 227 (Nov. 1972): 15–23.

Index